THE DEVELOPMENT PROCESS IN SMALL ISLAND STATES

Islands are coming under increasing environmental and social pressure, particularly as a result of the impact of tourism. In many ways, the small scale of these islands – almost enclosed systems – provides researchers with ideal cases in which to observe this process and test theory.

The Development Process in Small Island States focuses on the political security, tourism, gender issues, ecosystems, landscapes and economies of island communities. It encompasses islands at very different stages in the development process, identifying valuable common lessons and providing insights into the developmental issues particular to islands. With case studies drawn from the Caribbean, the Mediterranean and the Pacific, the book examines the position of islands as ecologically and economically vulnerable places.

Douglas Lockhart is Lecturer in Geography and **David Drakakis-Smith** is Professor of Development Studies, both at the University of Keele. **John Schembri** is Lecturer in Geography at the University of Malta.

THE DEVELOPMENT PROCESS IN SMALL ISLAND STATES

Edited by
Douglas G. Lockhart,
David Drakakis-Smith and
John Schembri

London and New York

First published in 1993
by Routledge
11 New Fetter Lane, London EC4P 4EE

Simultaneously published in the USA and Canada
by Routledge
29 West 35th Street, New York, NY 10001

© 1993 David Drakakis-Smith, Douglas Lockhart, John Schembri

Phototypeset in 10pt Garamond by
Mews Photosetting, Beckenham, Kent
Printed and bound in Great Britain
Mackays of Chatham PLC, Chatham, Kent

British Library Cataloguing in Publication Data

A catalogue record for this book is available from the British
Library.

Library of Congress Cataloging in Publication Data

The development process in small island states / edited by Douglas
G. Lockhart, David Drakakis-Smith, and John Schembri.
p. cm.
"Based on papers first presented at the Commonwealth
Geographical Bureau Conference on Small Island Development held
in Malta in March 1990"-Introd.
Includes bibliographical references and index.
ISBN 0-415-06984-X
1. States, Small-Economic conditions-Congresses. 2. Islands-
Economic conditions-Congresses. 3. Tourist trade-Congresses.
4. Economic development-Congresses. 5. Islands-Economic
conditions-Case studies-Congresses. I. Lockhart, Douglas, G.
II. Drakakis-Smith, D.W.   III. Schembri, John.
HC59.D462   1993
338.9'00914'2-dc20   92-26122
                                                      CIP

For Tom and Jean and in memory of
Bill and Jessica

CONTENTS

CONTENTS

FIGURES

TABLES

CONTRIBUTORS

Professor Richard Butler, Department of Geography, University of Western Ontario, London, Canada.

Dr Rajesh Chandra, Department of Geography, University of the South Pacific, Suva, Fiji.

Dr John Connell, Department of Geography, University of Sydney, Sydney, Australia.

Professor David Drakakis-Smith, Department of Geography, Keele University, Staffordshire, UK

Mr Israel Drori, Department of Sociology and Anthropology, Tel Aviv University, Tel Aviv, Israel.

Professor Henry Frendo, Department of History, University of Malta, Msida, Malta.

Mr Huw Jones, Department of Geography, University of Dundee, Dundee, UK.

Professor Russell King, Department of Geography, Trinity College, Dublin, Ireland.

Mr Edwin Lanfranco, Department of Biology, University of Malta, Msida, Malta.

Dr Anthony Lemon, School of Geography, University of Oxford, Oxford, UK.

Dr Douglas Lockhart, Department of Geography, Keele University, Staffordshire, UK.

Dr Janet Henshall Momsen, Department of Geography, University of California, Davis, California, USA.

LIST OF CONTRIBUTORS

Dr Robert Potter, Department of Geography, Royal Holloway and Bedford New College, Egham, Surrey, UK.

Mr John Schembri, Foundation Studies, University of Malta, Msida, Malta.

Professor Patrick Schembri, Department of Biology, University of Malta, Msida, Malta.

Dr Michael Sofer, Department of Geography, Tel Aviv University, Tel Aviv, Israel.

ACKNOWLEDGEMENTS

The editors wish to thank the Commonwealth Geographical Bureau and its Director, Professor John Parry, and also the Commonwealth Foundation, and particularly its Director In'oku Faletau, for providing financial support and the opportunity to organize the Conference on Small Island Development in Malta.

We also acknowledge the assistance of the Foundation for International Studies, under its Director, Professor S. Busuttil, who provided willing and efficient support in the preparation and organization of the conference. We wish to record generous grants from the Central Bank of Malta, the Bank of Valletta, the Malta Development Corporation and the British Council. Figure 13.1 is reproduced from *Scottish Geographical Magazine*, Volume 103(1), April 1987 by kind permission of the Royal Scottish Geographical Society.

Our thanks also go to Pauline Jones and May Bowers for typing the manuscript and to Andrew Lawrence and Mal Beech for cartographic and photographic assistance. Please note that the opinions expressed in the papers are those of the authors and not necessarily those of their employers.

The
Commonwealth
Foundation

Douglas Lockhart
David Drakakis-Smith
John Schembri

1

INTRODUCTION

Douglas G. Lockhart

This monograph examines a number of environmental and economic issues relating to island development. The chapters are based on papers first presented at the Commonwealth Geographical Bureau Conference on Small Island Development held in Malta in March 1990. The conference brought together about forty researchers from fifteen Commonwealth and four non-Commonwealth countries and was pitched at a broad level, not only to ensure the widest possible geographical coverage, but also to draw comprehensive material from islands at very different stages in the development process to see what lessons could be learned and exchanged.

Islands have long attracted the attention of geographers and researchers in cognate disciplines and, moreover, research has been spread over a range of economic, environmental and social issues. International conferences have been held at fairly regular intervals and most of the comparative studies of island development have originated at such meetings. One of the earliest initiatives to debate the human issues specific to smaller territories occurred in 1962, when the Institute of Commonwealth Studies in the University of London began a seminar series. Seminars were held at regular intervals over a two-year period and more than twenty papers demonstrated the extent to which many small countries shared common problems. About half of these papers were subsequently published in a volume edited by Burton Benedict (1967). In some ways this book set a pattern which many later works followed; chapters on general themes such as political, economic and demographic characteristics were followed by case studies of individual islands and smaller countries. However, the inclusion of Luxemburg, Swaziland and Tory Island (off the coast of Ireland) among the case studies gave the volume such a varied character that, not surprisingly, the conclusions which were drawn were tentative and probably account for the dearth of follow-up studies, at least until 1972 when a group of Sussex University researchers were joined by planners at a conference on small developing countries at the University of the West Indies in Barbados. The conference papers were subsequently published in a volume edited by Percy Selwyn in 1975. This book, and a discussion paper published three years later, concentrated on development policies, aid to small states and different types of dependent island economies.

The 1970s was, however, a period of dramatic changes with many island states achieving independence, and the process of forming regional organizations began to gather pace. Towards the end of the decade these changes were accompanied by heightened international interest in the problems of small island states. In particular, island states were given special status by the United Nations and by the United Nations Conference on Trade and Development (UNCTAD) and there was also a major change in Australian aid to Pacific and Indian Ocean island communities.

One consequence of all this interest was the choice of development problems of small islands as the topic for the 1979 Seminar of the Development Studies Centre of The Australian National University. The papers from this interdisciplinary seminar were published the following year (Shand 1980). In his introduction the editor sought to define smallness in terms of land area, population and the narrowness of the economic base. The book's chapters are grouped into eight sections that deal with primary activities; population and migration; trade and transport; economic stability; health and education; administration and politics; development strategies and aid and, lastly, an overview and assessment. Unusually only two chapters, Britton on tourism to Fiji and Fisk's analysis of development in Niue, discuss individual nations and the emphasis of the book falls heavily on more than twenty general chapters.

Island research received a major boost in the wake of the Falklands/Malvinas dispute in 1982. The crisis undoubtedly aroused interest in the study of small islands and one result was that post-crisis studies tended to give more weight to individual places.

Cohen's *African Islands and Enclaves* (1983), for example, was very much influenced by the conflict with chapters on the militarization of Diego Garcia and the structure of terror in Equatorial Guinea. More traditional concerns of social scientists, such as tourism, employment and economic diversification, feature in studies of the Canary Islands, The Seychelles and Mauritius. Such studies, and more particularly the wide-ranging review by Connell (1988), demonstrate that, far from being places of unchanging tranquility, many islands are in fact experiencing rapid transformation.

This point is also made with reference to the US invasion of Grenada in the introductory chapter of Clarke and Payne (1987) in their volume on *Politics, Security and Development in Small States*. This book emerged as a result of growing concern that current academic knowledge of the small states of the world was inadequate. It tries to answer the question 'How have small states coped with their smallness in the context of recent international economic and political trends?'. In their literature review the editors recognized that the division of subject material adopted by Benedict appeared to be most effective and in consequence their book can be divided into three sections. The first contained four surveys of the general situation encountered by small states, namely, the political, social, economic and security aspects of their existence. The second section comprised eight case studies of Commonwealth islands and enclaved states, each with a population of less than one million. The final section is reminiscent of the work of Selwyn and contains two broad perspectives on the problems of small islands taking contrasting academic and policy standpoints.

Finally, in this review of major texts on island states, a collection of essays on the theme of sustainable development and environmental management has been published as part of the United Nations Educational, Scientific and Cultural Organization's (UNESCO's) 'Man and the Biosphere' programme (Beller, D'Ayala and Hein 1990). Once again the book has its origin in a conference (held in 1986 in Puerto Rico). The introduction by Hess addresses a familiar theme; the problems emanating from smallness. The structure of the book followed the by now 'traditional' format with a further eight chapters dealing with general issues and fourteen case studies of individual islands or island groups. A notable feature was the wide difference in the island communities investigated and there were chapters on European islands such as Gozo, the Isle of Man and those off Brittany. The Indian Ocean was represented by Mauritius and there were several studies of the Pacific and Caribbean, while Bali, China's nearshore islands and Japan's remote islands each merited a chapter.

The wide-ranging observations and recommendations of the book reflect the interoceanic character of the conference. However, the editors identified a number of key common problems, notably 'distortions' on island economies, transport and accessibility, population change, unemployment, decisionmaking, natural resources and key economic sectors such as agriculture, tourism and industry. A range of possible strategies to ameliorate these problem sectors were summarized in a series of tables at the end of the volume.

In planning our conference, we were certainly guided by the trends identified in this overview of existing literature. As a result the primary aim was to bring together individuals who had undertaken research on a wide range of problems facing small islands in their attempts to develop. As noted earlier the conference was therefore pitched at a broad level, not only to ensure the widest possible geographical coverage but also to exchange information on the various stages of the development processs.

Malta was selected in part because the island had recently shown considerable interest in reviving its role within the Commonwealth. Moreover, as geography does not currently constitute a university degree programme in Malta, it seemed an excellent opportunity to show what geographers have done and can do. In addition, Malta itself could and did provide a living laboratory of some of the common problems faced by many small island states. In this context, the conference included two days of field visits in Malta and the smaller neighbouring island of Gozo. The conference embraced political, economic, social and environmental issues in island development.

The majority of participants addressed these issues in relation to particular case studies; however, we also invited Russell King and John Connell to prepare broader perspectives on island problems. Although revised versions of eight papers on environmental and economic themes have already been published (Lockhart and Drakakis-Smith 1991), this volume represents the principle conference publication bringing together the keynote addresses and a selection of substantive and revised papers. It is divided into two parts. The first addresses more general questions on island development, such as security, gender and basic needs, whilst the second half relates some of these issues to empirical studies of individual localities.

It is within this context of island studies that the contributions in this book should be placed. There are thirteen chapters in all. The first comprises the introduction to the volume; the next six are substantive surveys of aspects of island development and the remainder focus attention upon case studies of different development issues in widely-contrasting island locations. The authors are all leading researchers each of whom has wide-ranging experience of island communities.

In Chapter 2 Russell King, who has been researching processes such as emigration, tourism and dependency in various island communities in the Mediterranean for some twenty years, begins by discussing the attractiveness of islands to geographers and to other research workers. Next he attempts to define what is meant by an island in the context of size, population and political as well as physical relationships to neighbouring land masses. He considers the historic roles that islands have played and goes on to show how the colonial period has left its imprint in the fortifications and ecclesiastical architecture that are prominent features of many Mediterranean island landscapes. Another facet of island life is insufficient resources and a lack of employment opportunities to satisfy the aspirations of younger members of society. King describes how emigration has become a way of life throughout the Mediterranean and the Caribbean and he gives examples of formerly self-sufficient islands that have increasingly come to rely upon remittances from emigrants. Another recent change has been the growth of tourism and King assesses the uneven impact upon communities and landscapes. Drawing upon a number of recent studies he points to the need for greater harmonization of tourism development with the environment.

In Chapter 3 on political and security issues, Tony Lemon begins with a survey of the political status of island territories. He goes on to discuss internal political structures, and the relative stability of the governments of Commonwealth small island states, before finally examining the potential vulnerability of such countries to external pressures. His prognosis for the future of autonomous small states, written at a time of apparent easing global tensions, was relatively optimistic.

Janet Henshall Momsen in Chapter 4 deals with gender differences in the perception of environmental hazards in the Caribbean. She begins by identifying the main types of hazards that characterize small islands. Next she discusses land use change with reference to the feminization of agricultrure. Momsen argues that women in the Caribbean tend to be more active in the labour force than in most parts of the Third World, and this characteristic is particularly true of on-farm jobs. The field-work results from several islands demonstrates inter-island differences can be of greater significance than gender differences in attitude towards environmental problems.

The Caribbean is also the geographical focus for Richard Butler's contribution to the volume in Chapter 5. Butler reviews the literature on tourism on islands and offers a general model of development which provides a back cloth against which factors influencing development are discussed. Case studies from Antigua, the Cayman Islands and the Turks and Caicos Islands provide empirical evidence to support the theoretical model.

Robert Potter in Chapter 6 completes a trio of Caribbean studies by focusing on basic needs in the Windward Islands. More particularly, he examines the heavy dependence on imports and tourism and the comparative failure of policies which sought to achieve greater self-sufficiency and social progress in matters such as housing. Potter reviews a plethora of ill-co-ordinated initiatives. However, in highlighting the post-revolutionary experience of Grenada, the author notes with regret the brief economic renaissance based on grassroots development that was cut short by the military intervention of the United States.

In contrast, the final chapter in Part I by John Connell begins with a wide-ranging review of the development of natural resources, industry and tourism. Examples are drawn from islands in the Pacific, the Indian and the Atlantic Oceans. He emphasizes the difficulties of attaining economic growth through conventional strategies and Connell points to their survival through a mixture of remittances from emigrants and foreign aid. This much was evident in Potter's chapter. However, Connell shows that such trends are widespread throughout the Pacific and indeed in almost every island microstate.

The scale of analyses in Part II (Chapters 8–13) changes and here the contributors illustrate particular facets of island life drawing their conclusions from just one or two countries. Henry Frendo in Chapter 8 begins with a picture of the complex history of Mediterranean islands. He goes on to examine the colonial and post-colonial periods in Malta and Cyprus. In particular, his analysis highlights the relationship of Malta and Cyprus to the European mainland, and the ways in which British colonialism have blurred that relationship.

In Chapter 9 Huw Jones discusses demographic characteristics of small islands in general before commenting on recent demographic change in Mauritius, an island which thirty years ago suffered from grave problems of population pressure on limited resources. The eradication of malaria and an increasing birth-rate precipitated a population explosion in the 1950s and 1960s with an annual growth rate in population of over 3 per cent. Twenty years later the population growth rate had been cut to almost 1 per cent. He examines the factors which account for this spectacular fertility decline and demonstrates that growing female participation in employment and greater opportunities for secondary school education have promoted later marriage and smaller family size. Jones also emphasizes the role of family planning policy which has targeted rural as well as urban households before attention is given to temporal and local variations in fertility.

In Chapter 10, Michael Sofer and Israel Drori review socio-economic change in Fiji and Jamaica resulting from rural land settlement schemes. They present evidence to show how settlement projects in both islands mirror African experience in which the initial impetus for reform comes from state agencies and subsequently the initiative shifts to the settlers themselves. They discuss the characteristics of land settlement schemes on islands and then discuss two case studies, Narata village in Viti Levu, Fiji and Ebony Park, Jamaica. In both instances the schemes were state initiated and were later modified by the settlers themselves. The transformation involved adjustments of land use and labour reorganization and the authors describe in detail how problems have been tackled in contrasting circumstances.

In Chapter 11 Rajesh Chandra examines the industrialization process in Fiji. Given its status as a nascent industrializing state, Fiji provides important visions for other island states ambitious to transfer themselves into the new Singapore or Hong Kong. Chandra documents in detail how this process is heavily dependent on both foreign and state investment but is subject to distortion by localized events.

Douglas Lockhart (Chapter 12) discusses the problems of tourism development in Cyprus. He notes that in the post-war period the tourism industry has had to overcome enormous difficulties, the major problems being civil unrest in the late 1950s and again in 1963 and the military intervention by Turkey in 1974. Lockhart traces the historical development of tourism and shows that Cyprus shares many of the patterns characteristic of other Mediterranean islands. He then turns to the situation between 1974 and the present day. His central concern is to explain the contrast between the rapid growth of facilities in the Greek Cypriot South and the very mixed fortunes of the Turkish Cypriot northern area. Finally, he attempts to predict the prospects for tourism in the two communities.

Chapter 13 deals with the impact of development upon the changing landscape of the Maltese islands. Patrick Schembri and Edwin Lanfranco begin with an overview of the physical landscape illustrating the major geomorphological and climatic features and the typical habitats. Next they discuss the human impact upon that landscape, stressing the loss of countryside due to building development, quarrying and waste disposal problems. It is clear that these have contributed to the destruction of many ecologically sensitive areas. The authors also deal with soil erosion and grazing and the loss of specific kinds of habitat. They rightly point out that the Maltese natural environment has been subjected to a great variety of development pressures and that new approaches to countryside management are required in order to ensure an acceptable balance between development needs and the protection of the natural heritage.

REFERENCES

Beller, W., D'Ayala, P. and Hein, P. (eds) (1990) *Sustainable Development and Environmental Management of Small islands*, Paris, France: UNESCO.

Benedict, B. (ed.) (1967) *Problems of Small Territories*, London: Athlone Press.

Clarke, C. and Payne, T. (eds) (1987) *Politics, Security and Development in Small States*, London: Allen & Unwin.

Cohen, R. (ed.) (1983) *African Islands and Enclaves*, Beverly Hills and London: Sage.

Connell, J. (1988) 'Sovereignty and survival: island microstates in the Third World', Research Monograph no. 3, Department of Geography, University of Sydney.

Lockhart, D.G. and Drakakis-Smith, D.W. (eds) (1991) 'Environmental and economic issues in small island development', Institute of British Geographers, Developing Areas Research Group, Monograph no. 6.

Selwyn, P. (ed.) (1975) *Development Policy in Small Countries*, London: Croom Helm .

Selwyn, P. (1978) 'Small, poor and remote: islands at a geographical disadvantage', *Discussion Paper* 123, Institute of Development Studies, University of Sussex.

Shand, R.T. (ed.) (1980) *The Island States of the Pacific and Indian Oceans*, Australian National University Development Studies Centre Monograph no. 23, Canberra, Australia.

Part I

GENERAL PERSPECTIVES ON SMALL ISLAND DEVELOPMENT

2

THE GEOGRAPHICAL
FASCINATION OF ISLANDS

Russell King

The purpose of this chapter is to introduce some of the work that has been done on islands as a systematic geographic phenomenon. Islands have attracted the attention of scholars and writers from many disciplines. Geography, biology, ecology, anthropology, history and economics are perhaps the obvious ones, but we should not overlook the insights into island life and landscapes provided by poets and painters, many of whom have been drawn to the special island *ambience*. Islands spell romance and adventure and, following the genre of *Treasure Island*, *Robinson Crusoe* and *The Swiss Family Robinson*, have been the setting for many famous yarns.

Since previous conferences on islands – notably 'Islands '86' and 'Islands '88' – have concentrated on the Pacific and Caribbean islands of the world (see Beller 1986; Boutilier 1986; Chapman 1988), I shall take most of my examples in the pages which follow from the island-strewn sea that I know best, the Mediterranean. Given the profusion and diversity of islands in this region, such a choice would seem appropriate. The geographical position of the Mediterranean as the southern margin of Europe means that pressures for change involving processes such as emigration, tourism and dependency can be studied in a particularly clear fashion. My main concern, however, will be to describe the landscape changes that these processes provoke. Running through the latter part of this chapter will be the notion of cycles of landscape change on islands. Three such phases may be recognized. The first is the creation of an intensely-humanized landscape by the build-up of population over time, and especially during the nineteenth and early twentieth centuries. The interconnection between restricted space, dense population and an advanced and elaborate agriculture is a triadic theme repeated in many islands. The second phase is the abandonment and decay of this landscape through emigration and other forces. The third phase is the initiation of a new cycle of landscape trans- formation by the growth of tourism in the last few decades. This leads, finally, to interesting speculations on the role of islands in the future world economy and society.

THE ATTRACTION OF ISLANDS

An island is a most enticing form of land. Symbol of the eternal contest between land and water, islands are detached, self- contained entities whose boundaries are obvious; all other land divisions are more or less arbitrary. For those of artistic or poetic inclination, islands suggest mystery and adventure; they inspire and exalt. On an island, material values lose their despotic influence: one comes more directly in touch with the elemental – water, land, fire, vegetation, wildlife. Although each island naturally has its own personality, the unity of islands undoubtedly wields an influence over the character of the people who live upon them; life there promotes self- reliance, contentment, a sense of human scale (Wilstach 1926: 2–6).

For geographers, anthropologists, ecologists and biologists, islands hold a particular attraction, functioning as small-scale spatial laboratories where theories can be tested and processes observed in the setting of a semi-closed system. Geographers have long recognized this key role of islands. Yet in looking at the recent geographical literature on islands as a systematic phenomenon – as opposed to the very many case studies of individual islands – one notices a dearth of attempts to distil the essence of islands as a distinct geographic phenomenon. Notable exceptions to this general statement are John Connell's monograph on Third World island microstates, Stephen Royle's paper on the human geography of islands and a series of studies by the French geographer Emile Kolodny on Mediter- ranean islands (Connell 1988; Kolodny 1966, 1974, 1976; Royle 1989). Perhaps for the modern, hard-nosed geographer islands are too exotic, their study betokening a geographical dilettantism which detracts from academic credibility. Perhaps some geographers would concur with the economist Percy Selwyn (1980) who concludes that islands are not a particularly useful or meaningful unit for study.

I, and presumably the other authors of this book, would profoundly disagree with this view. The essential qualities of insularity – and often too of isolation – pose common problems for the thousands of inhabited islands the world over. Their existence as ready-made spatial models makes them ideal for the study of the interaction between population and territory which lies at the core of geography as a discipline.

If modern geographers have chosen to ignore this role, those of an earlier generation did not, for there is a rich vein of literature on the geographical significance of islands in the early twentieth century writings of French geographers like Paul Vidal de la Blache and Jean Brunhes, continued later by the great French geo-historian Fernand Braudel whose two-volume treatise on the Mediterranean contains a whole section of islands (Braudel 1972: 148–67). Also very noteworthy is the work of Ellen Churchill Semple whose book *Influences of the Geographic Environment* contains a long chapter on islands (Semple 1911: 409–72).

What does this early work tell us about islands? Space allows only a rather crude and selective summary of the richness of detail and insight. The following topics will be dealt with: island definitions and concepts of insularity; historical roles that islands have played – a shifting dialectic between isolation and focalism; and the theme of cycles of landscape change as stimulated by three consecutively-phased processes – population pressure, emigration and tourism.

WHAT IS AN ISLAND?

First, of course, there is the question of definition, more complex than might appear at first sight. Some islands are surrounded by water only at high tide, or become islands only at low tide, being below water at high tide. Processes of coastal deposition may eventually link an island to the adjacent mainland; conversely erosion or a storm may break such a link where it exists already. Human agency can destroy or create insularity by the building of a bridge or the excavation of a channel through an isthmus. According to the Vikings, an island was defined as such only when a ship with its rudder in place could pass between it and the mainland. History does not record how many Viking longships were wrecked testing this definition, but it does tell us the tale of Magnus Bareleg who, having gained the right by treaty to possess all the islands off the west coast of Scotland, in 1098 had his men haul his ship across the peninsula of Kintyre and thus laid claim to it (Grant 1982:151).

A further definitional problem concerns scale, and thresholds of minimum and maximum size. This is a topic of potentially endless debate, for there is no widely-agreed set of limits (the early geographers were wisely silent on this issue, sensing the futility of extended argument). A few anecdotal examples will serve to illustrate the point. How small can an island be before it is classed as a lump of rock, bank of sand or reef of coral? The Scottish Census of 1861 declared that an island was an area of land surrounded by water inhabited by man where at least one sheep could graze. Smallness of islands is not just a matter of definitional hair-splitting, for it is co-involved with the ownership of the sea and, therefore, of fishing and mineral resources. The International Convention of the Law of the Sea specifies that rocks which cannot sustain habitation or economic life of their own shall have no exclusive economic zone or continental shelf; which declaration prompted a British subject to spend a month in 1985 encamped on the pinnacle of Rockall 320km west of the Outer Hebrides (Hache 1987:88).

At the other end of the size continuum, how big does an island have to be to lose its insular character? Australia, 7.8 million km^2 and 14 million inhabitants, is classed as a continent. Greenland, 2.2 million km^2 and 50,000 inhabitants, is recognized by some as the world's largest island, by others as its smallest continent. If insularity is to be linked conceptually to problems of economic disadvantage and marginality, how about Britain and Japan, island nations which have hardly suffered through their insularity: yet Royle (1989) includes them in his 'world list' of islands. Doumenge (1985a) tries to invoke climate as a criterion for 'islandness', a highly dubious exercise. He suggests that when an island has an emerged volume large enough to create its own climatic effect (he nominates 20,000 km^2 of surface and a mountain mass of over 1,000 metres mean altitude), it enters the continental category. Iceland, Tasmania, Taiwan, Madagascar (among many others) are thus to be classed as 'continental islands' and not 'true' islands. The danger of physical determinism is present here, for the characteristics of island peoples depend less on climate than on a whole host of other factors such as density and distribution of population, geographical location (for example, with respect to adjacent land masses), cultural background, seafaring tradition and the general level of economic and political independence (cf. Hache 1987:88).

Elsewhere in the literature there is an attempt to sidestep the problem of 'how big can an island be' by narrowing the discussion down to issues concerning small islands – an approach followed by this book. Suggestions for an upper limit for small islands include 10,000 km² and 500,000 residents (Beller 1986:9), and 13,000 km² and 1 million people (Dolman 1985:40). Within the context of the Mediterranean such thresholds are not particularly helpful since they introduce an artificial distinction between, on the one hand, Sicily, Sardinia and Cyprus which to varying extents exceed the limits specified above, and qualitatively comparable islands like Majorca, Corsica and Crete which do not.

Finally, in this introductory section, it must be pointed out that islands exist in a variety of political and administrative contexts. Some, like Malta, are politically-independent microstates. Others are islands belonging to larger metropolitan countries, both near (for example, the Isle of Wight to Britain) and far (for example, Guadeloupe to France). Other islands are part of archipelagos that have their own internal cohesion, embracing either entire states (for example, the Philippines) or significant parts of countries (for example, the Greek islands of the Aegean Sea). Depending on their particular features of political geography and location, islands may suffer from multiple isolation and peripherality. The small Mediterranean island of Alicudi, for instance, is marginal to the Aeolian archipelago and to its main island of Lipari, but the Aeolian islands are themselves a peripheral part of Sicily, which in turn is peripheral to the Italian mainland.

STRETCHING THE CONCEPT

The notion of insularity has a number of geographic expressions. Brunhes (1920:52) discusses islands of the sea as but one variant of a range of geographical types of 'islands of humanity' – oases in the desert, high mountain valleys, and islands of cultivation and settlement in the equatorial or boreal forests. As the isolating factor the sea thus has a number of distinct analogies: sand in the desert, the barrier of high mountainous relief, or the extensive and often impenetrable forest. Later in the same book Brunhes gives detailed accounts of specific examples: the Suf and Mzab oases of the Sahara, the isolated intermontane basins of the Bolivian Andes, and the Balearic islands of Majorca and Minorca (Brunhes 1920: 415–513).

In his intriguing account of islands in the Mediterranean Braudel, too, extends the discussion to 'islands that the sea does not surround' (1972:160–1). Although he acknowledges that he is taking great liberties with the concept of insularity, he notes historical similarities in the lives of islands and peninsulas. Indeed, according to him the whole Mediterranean region can be likened to a set of quasi-islands isolated by sea and mountain, yet trying to make contact with one another. Spain has been described as a *plusqu'île*, 'more than an island', to emphasize its inaccessibility and original character. The Maghreb, too, Braudel notes, has been termed an island, less part of the African continent than an isolated milieu firmly bounded by the Atlantic, the Mediterranean, and the Sahara Desert.

Third, the geographer's physical concept of islands surrounded by sea (or sand, or whatever) has a sociological parallel in the notion of 'social islands', groups of people who are inward-looking and cut off from the rest of society by externally-imposed or self-inflicted social boundaries (Pitt 1980). Although separate, social islands are, however, related to the wider social context (the sea) and to larger societies (the 'mainlands' of mass capitalism or socialism). As Pitt (1980:1052) goes on to point out, the social category of islands shades into the peninsula, the social island with tenuous connections to the mainland, the isthmus being the nature of the narrow link to the main society. Further sociological/

geomorphological parallels may also be suggested: 'continental' islands (large populations, long separated from the 'mainland' society) are different from 'diastrophic' islands (groups isolated by the rising and falling of social movements around them), or from 'volcanic' islands (social groups created by sudden, perhaps violent, means), or 'atolls' (quietly built from the gradual accretion of custom and tradition). A physical archipelago of islands has parallels in the Jewish ghettoes of many European cities before the war, or in the social mosaic of different immigrant or slum groups in the same city – cf. Wakefield's *Island in the City* (1977).

HISTORICAL ROLES THAT ISLANDS HAVE PLAYED

Islands have fulfilled a number of distinctive roles in the evolution of cultural, geographical and developmental patterns. Two contrasting roles can be immediately counterposed: on the one hand the obvious link with isolation, on the other a nodal role as 'centres for the gathering and fusion of peoples' (Vidal de la Blache, 1926: 157).

The first of these is the most typical role, and one that has been expressed in several ways. For instance, at both the biological and ethnographic level, isolation has produced the survival of endemic and archaic forms of life. Biological endemism is illustrated by the tiny Shetland pony or Sardinian donkey. The dwarf species of elephant and hippopotamus surviving on Malta until about 8,000 years ago are another example. Environmental controls – poor food, a restricted spatial environment, excessive inbreeding – have been partly responsible. The Maltese elephant, only one metre tall, was a fraction of the size of its African and Asian cousins. Species uniqueness on islands was also stressed by Darwin: out of twenty-six species of land birds described by him in the Galapagos islands, all but one were unique to the archipelago.

In the human realm, islands have often become cultural backwaters where ancient traditions survive, whereas elsewhere the march of 'progress' has swept these features away. The role of islands as ethnographic museums has attracted to them many anthropologists and folklorists to study customs and artefacts. Linguists may be interested in the archaic forms of speech conserved on islands – an example is the Sardinian dialect, closer to Latin than any other Romance language. The inward-looking nature of some islands is often exacerbated by a lack of coastal settlement: such is the case with Corsica and Sardinia, for example, most of whose villages and towns are sited well away from the coast. Islands often shelter the last traces of dying civilizations or of social, political and religious movements. Anglesey was the refuge of the Druids who fled from the Romans; Ceylon/Sri Lanka became the refuge of Buddhism driven out of India; and Formosa/ Taiwan was the refuge of the Ming dynasty during the political troubles of the seventeenth century, just as three centuries later it was the final bastion of Chinese nationalism against mainland communism (Perpillou 1966:18–19).

Islands also reflect their isolationist role in the way that they are used as places of exile for deposed figureheads and criminals, or as locations for monasteries or colonies of artists. Napoleon, an islander all along (he was born on Corsica), was banished first to Elba, then to St Helena. Mussolini perfected the practice of island exile, and the modern Italian state still sends *mafiosi* to insular exile on small barren islands like Linosa (off the south coast of Sicily) and Asinara (off the north coast of Sardinia). Recently, media attention has been drawn to the island of Léros where the Greek government has dumped hundreds of mentally-ill people in a colony in widely-condemned inhuman conditions.

The second general historical role of islands is that they can, under certain conditions, act as foci for advanced civilizations and cultures. In a non-urban setting, population growth within a fixed perimeter can stimulate remarkably-elaborate agricultural landscapes, as the rice terraces of eastern Java, Bali and Lombok illustrate. Similarly intensive agricultural systems are found on many other islands in different parts of the world. Such population pressure may ultimately lead to emigration, but not before the landscape has been transformed to a degree not found on the mainland. Where civilization is essentially maritime, islands can become important nodal centres, as happened in the Mediterranean under various colonial and thalassocratic regimes using so-called *routes des îles* (Pryor 1988:91). Semple (1911:266) noted that islandless seas attained a later historical development than those 'whose expanse is rendered less forbidding by hospitable fragments of land'. The contrast between the northern and southern shores of the Mediterranean would seem to bear out this assertion.

Islands have often been the first places to be incorporated into expanding empires, forming staging posts en route to further conquests. Thus the Phoenicians expanded first to Cyprus, the Greeks to Sicily, the Etruscans to Corsica, and the Romans to Sardinia. The advance of Spain to the Canary Islands was 'the drowsy prologue to the brilliant drama of American discovery' (Semple 1911:428). Other islands' strategic location on trade routes explains their commercial importance, as exemplified in the cases of Zanzibar, Singapore and Hong Kong. Imperial ambitions linked to economic exploitation led to the Portuguese occupying a string of islands on key routes around the world: Madeira, the Cape Verde Islands, and the Azores on the Atlantic trade routes and, further afield Ceylon, Malacca and the Moluccas. Islands have also played an interesting role in the dissemination of agricultural crops, as the example of sugar illustrates. Brought from Egypt to Cyprus in the tenth century, it then diffused westwards through a chain of islands (Sicily, Madeira, Azores, Canaries, Cape Verde) to reach the Caribbean Islands and the Americas (Perpillou 1966:483).

As a number of examples from the Mediterranean testify, the dual role of geographical isolation and an early incorporation into cultural impulses has produced some striking examples of advanced civilizations existing in quite isolated insular locations. To name but a few, we may note the Minoan civilization in Crete, still enigmatic in many respects, not least in the circumstances which led to its sudden demise, and the equally mysterious megalithic cultures of Malta, Corsica, Sardinia and the Balearic Isles. In contrast to their past greatness, most of these places have been economically and culturally marginal in the modern historical era. Thus islands may play different historical roles at different periods. Semple (1911:412-17) recalls the contrasts in the early and later histories of various islands or island groups such as the Canaries, the Azores, Mauritius and Hawaii: 'now a lonely, half-inhabited waste, now a busy mart or teeming way-station'. Easter Island provides perhaps the clearest illustration of this contrast. Once it was densely populated and completely tilled by a people who had achieved singular progress in farming, religion, sculpture and hieroglyphic writing. But twentieth-century Easter Island shows only 'abandoned fields, the silent monuments of its gigantic stone idols, and the shrunken remnant of a deteriorated people'.

Finally, islands have often been fought over, usually for their strategic location or, as in the case of the Falklands, for their symbolism of empire. Braudel (1972:161) writes that the history of the Mediterranean islands is like an enlarged photograph of the history of the Mediterranean region as a whole. As the least-weighty fragments of land, history has tossed them hither and thither. Sardinia in the Middle Ages was contested by both Pisa and Genoa, their solicitude accounted for by Sardinian gold. Catalan expansion also found a foothold here. Crete and Cyprus fell under the yoke of both Venice and the Ottoman Empire, as well as many other colonizing powers at different periods of history. During the long struggle between Islam and Christianity for maritime supremacy in the Mediterranean, control of the islands which dominated the 'trunk routes' of the sea was crucial (Pryor 1988:101). Later, Gibraltar, Malta and Cyprus formed a vital strategy for British control of the Mediterranean and the routes beyond Suez.

All these colonial occupations have left their indelible imprint on the landscape. In Sardinia there are Pisan Romanesque churches and Genoese watch-towers; Catalan is still spoken in the town of Alghero. At Rethymnon (Crete) a mosque is found within the Venetian ramparts of the citadel, whilst at Famagusta (Cyprus) minarets sprout from the old Lusignan Gothic cathedral. The Moorish occupation of Sicily has left its mark forever in the place-names and dialects of the island. In Cyprus the British military base at Episkopi contains leafy suburbs that could have been lifted straight from Surbiton or Orpington.

Just as islands have often been the first to be incorporated into empires because of their ease of conquest and strategic importance, so they are often the last to be decolonized. Small island states are the most numerous element of the Commonwealth: of forty-eight member countries, twenty-six have populations of less than 1 million people, and all but seven of these are islands (Ramphal 1988:3). Semple wrote (1911:430-1) that island fragments of former empires were found everywhere, seemingly innocuously scattered around the world: for instance the British, French, Dutch and Danish holdings in the lesser Antilles, island monuments to lost continental domains. The retention of small islands in an era of decolonization stems partly from their strategic value and partly from the islands' need for economic support from the imperial power.

ISLANDS AND EMIGRATION

The build-up of island populations has limits set within a particular economic and technological system. When this limit is reached, one of three outcomes occurs: malnutrition or starvation results; help has to be sought from outside, for example, in the form of food shipments or welfare payments; or, most likely, emigration takes place. The smaller the insular space, the earlier is the point of saturation reached. Over-burdened with excess population, islands the world over have been fountains of emigration. Virtually all the islands of the Mediterranean and the Caribbean have exported their populations and there are many classic cases of island diasporas recorded in the geographical and anthropological literature. For the Mediterranean, Braudel (1972:158–60) describes Corsica as the island of emigrants *par excellence*. In spite of being an island of mountain shepherds (unlike many other islands where a tradition of seafaring led on naturally to migration), Corsican communities existed in many of the great Mediterranean seaports by the sixteenth century. Today there are more Corsicans on the French mainland, in Marseilles and Paris, than there are on Corsica itself.

Emigration may become institutionalized as part of island society, and necessary for its stable survival. For the Aran islands off the West coast of Ireland, Gailey (1959) concluded that the status quo could only be maintained through continued emigration. Many islands, however, have suffered massive population decline as a result of emigration. This may happen, for instance, when emigration snowballs through a 'contagion effect', or when the economic base of an island's livelihood is exhausted or wiped out, or when improved communications allow an easy exodus to explore economic opportunities elsewhere. Thus, within half a century, a wave of heavy emigration reduced the population of Ithaca, an island off the west coast of Greece, from 11,409 (1891) to 5,877 (1951).

The sequence of events on Ithaca may be taken as fairly typical of many Mediterranean islands (Lowenthal and Comitas 1962). Until the early twentieth century Ithaca managed to be practically self-sufficient, its rugged and arid terrains sustaining sufficient wheat, olives, vines, vegetables and animal products to support its inhabitants. When the population increased beyond the carrying capacity of the land, Ithaca's long seafaring tradition staved off economic crisis. Young men took to the sea to provide for their families, and Ithaca became a peasant society supported by sailors who returned regularly to participate in the life back home. After 1910, however, the temporary migration of sailors began to lead to more permanent settlement abroad as opportunities presented themselves in the cities of the United States and Australia. Meanwhile, the impact on Ithaca itself was being felt as the metamorphosis from a peasant society to a remittance society took place. A shrinking population was becoming demographically unbalanced with a surplus of females and fewer and later marriages. Labour shortages in farming led to land abandonment and a halving of the value of agricultural production. Only olives, a crop requiring minimal labour, maintained their output. Remittances from Ithacans abroad bridged the economic gap, but only at the expense of diminishing self-reliance and growing imports.

In landscape terms the most obvious effects of emigration from islands are the abandonment of land and the desertion of settlements. Sometimes the decline of farming takes the form of a shift to a less labour-intensive regime – pasture instead of tillage. As depopulation continues, the per capita cost of services rises, and the burden of maintaining schools, roads and other community facilities falls increasingly on outsiders, in particular the government of the mainland. Schools, shops, churches and other facilities close down. Because of the demographic selectivity of the emigration process, those who remain – the very young, the old, the infirm, the unskilled – produce less than the emigrants and need more. Hence welfare and remittances loom larger and larger in the economic accounts of island societies.

Detailed studies of landscape degradation in the Aeolian islands reveal some interesting micropatterns in the process of land abandonment (Buffoni 1987; King and Young 1979). Generally it is the upper and more remote terraces which are withdrawn from agriculture first; these are the plots which are least accessible and furthest from the villages and hamlets, most of which are located either at sea-level or at intermediate levels on benches and saddles which carry the few roads and tracks. Other plot-by-plot contrasts are the result of the impact of emigration on the tenure pattern: some emigrants sell their land to relatives and neighbours; those who do not, leave their land to be reclaimed by the *maquis*. Crumbling terrace walls and locked-up decaying dwellings complete the picture of landscape dereliction. On Salina, where the processes of Aeolian rural decay are to be seen in their sharpest focus, emigration was stimulated most of all by the phylloxera epidemic which destroyed the island's important vineyards in the 1890s. The vines, which had been planted to produce malmsey wine largely for the English market, had become a virtual monoculture upon which the population had become over-dependent. Although some replanting took place, emigration was widespread and population decline inexorable: from 4,907 in 1871 to 2,193 in 1971, by which time 39 per cent of the dwellings on the island were vacant (others, abandoned earlier, had simply disappeared into the landscape).

The above pattern of demographic and landscape decline has characterized virtually all Mediterranean islands and island groups in the recent past. Of course, there are differences in the rate and timing of the outflow. Some islands like Corsica and Salina have been steadily losing people for a century or more. In others the decline is more recent and perhaps more sudden. The population of Kastellorizo (Megísti), off the coast of Turkey, fell from 9,000 in 1910 to 2,750 in 1922, a dramatic decline due to the closing-off of sponge-fishing in Libyan waters and of access to agricultural land on the Turkish mainland (Kolodny 1966:16). The main exception to the general trend of net emigration is to be found in the Balearics, where economic growth following the model of tourism has led to positive migratory balances over all recent inter-censal periods (Kolodny 1976).

Outside the Mediterranean, emigration has had a similar impact on agriculture, leading to a general decline in the use of marginal and distant land (Connell 1988:38). Fiji recorded a diminishing cultivated area by the 1950s and in several Caribbean islands as much as a third of the agricultural land now lies idle. Migration and remittances have accelerated the disintensification of the traditional multi-crop agricultural system which in many Third World islands had already been deformed by plantation agriculture. In classic island plantation economies, such as Cape Verde, landlessness remains widespread despite considerable outmigration (Meintel 1983).

In many islands emigration is not an irreversible process; people do not necessarily leave never to return. The nostalgic pull of an intimate insular homeland brings many emigrant sons and daughters back. Many Mediterranean (and other) islands have established traditions of seasonal and temporary migration which channel remittances and savings back to support the island economy. The capital inflow can help to transform the landscape, usually through investment in property – agricultural land or, more often, new buildings. The case of Malta shows how the emigration experience which is virtually endemic to Mediterranean islands has made a bold imprint on the built environment, with spacious new 'emigrant houses' in every village as well as splendid new parish churches financed by emigrants. On Gozo, especially, the fashion is for returned emigrants to adorn their new villas with symbols (stone kangaroos, eagles, and so on) and name-plates ('Sydney House', 'God Bless America') which reflect their personal emigration history (King and Strachan 1980).

ISLANDS AND TOURISM

As well as returning migrants, there are other recent processes affecting islands which contribute to a new cycle of landscape change based on coastal construction activity. Such new processes stem from islands' attractiveness as tourist spots. Although this mass discovery of a new 'use value' of islands is relatively recent, dating back in the main to the 1960s, there are earlier antecedents. Nearly two thousand years ago the Romans built their holiday villas on the Isle of Capri, and eighty years ago Semple (1911:452–3) wrote also of the islands of the Neapolitan Gulf (Capri, Ischia, Procida) which have 'peaks which attract rain by their altitude and visitors by their beauty, and a mild climate delightful in winter and in summer'.

The touristic attractions of islands are too well known to list in all but the most summary of forms: they include peace and quiet, unpolluted air, clean sea, empty beaches, friendly local people and wonderful landscapes. Some islands also have an appealing architectural and cultural heritage. The same sense of isolation which operates in a negative way for islanders, adding to their physical and social remoteness and to their transport costs, also acts as a positive factor encouraging tourists to come and 'get away from it all'. The difference of course is that a few weeks of escapism hardly equates with a lifetime of marginality and limited opportunity.

Many of the readers of this book will be familiar with the tourism cycle of landscape change through their own holiday experiences. Once again the Mediterranean, Europe's holiday playground, provides exemplary case-study material, although similar landscape changes are to be observed in many islands of the Caribbean, Indian and Pacific Oceans and in the islands of the 'Mediterranean Atlantic' – Madeira, the Canaries and the Azores. Initially the impact may be small – a few 'tavernas' and some modest accommodation offered to visitors. Then the establishment of regular ferry links (including car ferries to the bigger islands like Corsica, Sardinia and Crete), followed by the building of airports for charter flights (Corsica and Sardinia have no fewer than seven charter flight airports between them), heralds the arrival of mass tourism and the inflow of multinational capital to develop tourist resorts. Big hotels and holiday complexes start to transform the built environment of coastal areas: formerly sleepy fishing villages are swamped by new construction activity, and previously deserted stretches of coast witness new settlement growth, generally focused around one or more big hotels or estates of villas. Mellieha on the north coast of Malta provides a good example of how the building of space-consuming villas on prominent clifftops destroys pristine coastal scenery. On the Tunisian island of Djerba beach-front villas with their outward-facing sun patios and their broad access roads for motor traffic introduce a marked contrast with the traditional spatial pattern of life and settlement based on

27

tightly-clustered, inward-looking dwellings separated by a maze of narrow, winding alleys. On Djerba, too, uncontrolled building of tourist developments on sand dunes has upset dune ecology and resulted in rapid coastal erosion (Miossec 1976). The clash of imported villa styles with vernacular architecture can also be clearly observed in the Balearic Islands where, in the mushrooming of places like San Antonio (Ibiza) and Palma (Majorca), one arrives at the fully mature 'tourist city'. The privatization of island space by the creation of tourist enclaves into which the locals may not enter (except under certain conditions, for example, in the role of waiters, chamber-maids, gardeners, and so on) completes not only the cycle of landscape transformation, but also the neo-colonial take-over of the island's economy and resources to service outside metropolitan interests.

Naturally we are here entering the realm of value judgements and there is a danger of imposing external academic and probably middle-class values on to a situation in which we ourselves are not directly involved, except perhaps as temporary consumers. This is the classic planner's dilemma. To see tourism as a capitalistic device for the systematic destruction of everything that is beautiful in the island world is to gloss over real questions of economic growth and linkage to other sectors of the island economy, and to ignore the views of the islanders themselves. Majorca and Ibiza have been widely criticized for allowing excessive tourist development and the charge is largely justified: much of the development has been hasty, uninspired, even tasteless. But one should not be too elitist: it happened in response to demand and clearly many people still enjoy visiting these islands which have become one of the richest parts of Spain.

The economic importance of tourism to small islands in the modern world economy is not to be underestimated: in at least half of the island microstates (with populations of less than 1 million) in the developing world, gross receipts from tourism are larger than all visible exports put together (Dommen and Hein 1985:163). Of course, the import content of the tourist industry tends to be very high, since tourists often want much the same things as they have at home. As Dommen and Hein go on to point out, this applies not only to cornflakes but also to the facilities and management style of the hotel, as well as to levels of energy demanded (for air-conditioning, hire-cars, and so on). On the other hand, the tourist industry can provide the basic level of demand to make economic the provision of certain goods and services (water, elecricity) to the population as a whole. Tourism can also stimulate other, related, economic activities. Some of these are obvious, such as construction and the servicing of the entire tourist apparatus (lifts, swimming pools, hire-cars, and so on). In the case of more independent economic activities such as fishing and farming, the picture varies from one island to another. In Ibiza, and to a lesser extent on other Mediterranean islands, local agriculture is stimulated and market gardening thrives. Elsewhere tourism has actually retarded local agriculture as workers are enticed from the land to serve as waiters, porters, taxi-drivers, and so on. The Bahamas and Bermuda, islands which are outstandingly successful on the tourism front, import virtually all their food (Dolman 1985:47).

Islands seeking tourism as a panacea are, however, operating in a highly-competitive and volatile market. Whilst it is true that there has been spectacular growth in world tourism in recent decades, this growth has not been consistent, either in space or over time. Economic recession, changing exchange rates and the fluctuating price of energy (especially aviation fuel) cause booms and slumps in numbers of tourists which can be particularly devastating when applied to a small island economy overly dependent on tourism as a major economic platform. Tourism can bring rich rewards to small islands, but it is a 'fragile dependency' (Wilkinson 1987). Such fragility is often exacerbated by an overreliance on one nation as the main supplier of tourists. Malta's reliance on British tourists is a case in point: the collapse of tourists visiting Malta from 730,000 in 1980 to 480,000 in 1984 was due entirely to the drop in arrivals from Britain, non-UK visitors holding steady at around 200,000 per year (Lockhart and Ashton 1987). The recovery to a total of 746,000 in 1987 and then to more than 800,000 in 1989 was due first to a resurgence of British interest (in 1987), and subsequently to growth in tourists from other countries such as Italy and West Germany. In the small islands of the Caribbean such as Antigua and St Lucia tourism has been likened to a plantation industry (Weaver 1988). The importance of foreign capital, entrepreneurial input and managerial control, recruiting of cheap unskilled local labour, seasonality and reliance on a narrow external market, are all features which make island tourism similar to the plantation agricultural system. In many tropical islands the decline of plantation cropping has coincided with the rise of 'plantation tourism', thus ensuring the continuity of the plantation model.

According to Beller (1986:31), tourism as practiced on small islands produces generally beneficial economic results, but mixed social and environmental impacts. Undesirable sociocultural spillovers include: cheap commercialization that tends to destroy the measured pace of island life; rising crime rates, related especially to theft, drugs and prostitution; the invention of a spurious 'folk culture' which becomes adopted to satisfy tourists' curiosity; and the admiration of many things foreign which works to weaken local culture and creativity. These problems have perhaps been most acutely felt in the Caribbean.

Environmental problems of tourism have to do with the saturation of tourist numbers and their spatial concentration around the coast. The predominant style of tourism on islands is based almost entirely on the attractions of climate and beaches. This style blurs the geographical differences that exist between islands and forces each island to compete world-wide with all other 'tourist' islands. Concentration on beach tourism may constrain recreational opportunities for local people, leading, as in Barbados, to a 'windows on the sea' movement to retain for the local population at least some visual access to the water (Beller 1986:30). Beach tourism also neglects the potential attractions of an island's interior, where one may find historical, cultural and scenic resources of real interest such as rainforests, volcanoes and archaeological sites. In larger islands such as Cyprus, hill stations (to escape the summer heat) and even skiing are on offer.

Tourism depends on environmental quality more than any other economic activity. Although, in principle, natural features of landscape, vegetation, wildlife, clean sea, and so on, should be protected in order to safeguard the future of tourism (not to mention the future of the island ecosystem as a whole), in practice tourist development implies consumption of land and vegetation, pollution of the sea, urbanization of the coast and the exploitation of limited insular resources. The potential effects of environmental degradation due to excessive and uncontrolled tourist development can ultimately lead to the destruction of the very environment the tourists come to enjoy. Critics of mass tourism's propensity to overload infrastructure, destroy environmental resources and swamp island culture, increasingly point to the attractiveness of low-density styles of tourism based on affluent, low-volume, long-stay and repeat visitors. Environmentally- and culturally-based specialty types of tourism like diving schools, wildlife tours, scientific tourism and retirement tourism are some individual examples (McElroy *et al*. 1987; Wace 1980). An ecologically-based reshaping of islands' tourist potential is stressed in several papers in a recent special issue of the journal *Ekistics* on small islands (see Buffoni 1987; Gomez Sal *et al*. 1987; McElroy *et al*. 1987; Papaioannou 1987). In pursuing an ecodevelopmental strategy these proposals for specific islands like Crete, Salina, Gomera, Mauritius and Hawaii argue for extending the planning horizon beyond the next generation, holding the pressures from tourism in check in order to create sustainable development for islands in the longer term – a point I shall return to in the concluding discussion. As an example, Buffoni's plan for Salina involves the creation of a multi-purpose nature park comprising the island's marine resources as well as its unique terraced landscape. Such a park would function as the 'fulcrum of the social and collective management' of the island for generations to come; it would foster the appreciation, conservation and restoration of the island's agricultural landscape instead of a destructive exploitation of purely coastal areas coupled with neglectful decay of the interior (Buffoni 1987).

CONCLUSION

The above remarks on the ecodevelopmental approach to island planning and to balancing human pressures deriving from tourism with the continuing viability of the natural resource base bring us back to our much earlier discussion of early twentieth-century geographers' views of islands. Jean Brunhes (1920:499) pointed out that islands called forth the first true regional monographs; based on observation of the essential facts, such comprehensive studies of 'little wholes of humanity' served as an effective introduction to the regional method, leading on both to the study of larger and less-easily defined units and to the more detailed examination of problems of ecology and of historical, social, economic and political geography.

According to Hache (1987:89–90) three schools of thought can be observed in the study of islands and island problems. For some, the extreme variety of island situations precludes their study as a specific phenomenon. As we saw earlier, this is the opinion of Selwyn (1980) who states that neither economic structure nor social trends can be usefully studied in an island context. Such a view is echoed by Murray (1984) who notes that, in the case of islands, smallness is far more relevant than 'islandness'. The opposite school of thought states that isolation and insularity are the key characteristics of islands. This view is best exemplified in the work of Brunhes (1920), Braudel (1972) and Pitt (1980), all of whom extend the notion of islandness to other realms beyond the simple geographical definition of an island. The third approach avoids the metaphysical controversy of 'what islands are' and concentrates instead on means of studying them as vehicles to elucidate particular methodological and empirical approaches such as the core-periphery model (Alexander 1980), endemism (Doumenge 1985a) or the gravity model (Doumenge 1985b). My own sympathies lie with the second and third of these schools of thought, but not with the first.

Contemporary regional and planning theory as applied to islands wavers between the ideological precepts of core-periphery dependence and neo-classical comparative advantage. Both stress the particularities of isolation and structural weaknesses of island economies, but both tend to overlook the important linkage elements of island ecosystems (Coccossis 1987). The ecosystem uniqueness of islands also lends weight to the second school of thought discussed above.

These is a strong ecological theme running through much of the research on islands, particularly that carried out from a biogeographical perspective and published in a number of standard texts in the 1960s and 1970s (Carlquist 1974; Fosberg 1965; Gorman 1979; Lack 1977; MacArthur and Wilson 1967). More recently, increasing pressure from rampant speculative development associated with tourism and urbanization, and a neglect of the deteriorating rural hinterland, are re-focusing

32

attention on to ecological principles, to questions of spatial balance, and to the essential interconnectedness of all elements of the island scene. Many writers are convinced that for most islands the opportunities for development are closely intertwined with the preservation and improvement of their environmental resources which have been damaged in the past. As McElroy *et al.* (1987:93) put it:

> Centuries of deforestation, erosive plantation monoculture, marine exploitation, natural disasters, and policy neglect have produced the progressive loss of renewable resources, diminished biological productivity and diversity, and the abandonment of traditional resource husbandry. The early institutionalisation of chronic and large-scale emigration from these fragile, degraded insular environments is a bleak testimony to the unsustainable legacy of the colonial economy. For islands undergoing the contemporary transition from boom-bust agricultural staples to more durable tourism and export manufacturing, landscapes and coastlines are being visibly altered by widespread urbanisation and the intrusive impact of hotel, marina, airport and refinery construction.

Thus the ecological approach – the effort to combine development and conservation within a holistic framework – becomes central to the problematic of island planning (Coccossis 1987:84).

Island ecosystems display certain key differences which distinguish them from continental ecosystems or even from small, non-insular regional ecosystems. These distinguishing features, some of which can be listed as follows, embody powerful implications for the type and intensity of development which can be contemplated in an insular milieu. First, islands seem to be ideal examples of closed systems; however, when such systems are opened to outside influence they become unstable and fragile. Second, island ecosystems are characterized by reduced inter-species competition, archaic forms of life due to isolation, and genetic diversity: these combine into a delicate and intricate balance which, again, exhibits vulnerability if upset. Third, similar mechanisms are observable in the social realm. Separation and isolation have the general effect of supporting traditional customs and other socio-cultural traits, fostering a sense of identity and belonging. However, when these patterns are disturbed, for example, by mass tourism, reactions are sudden and sweeping. Fourth, the island economy presents a similar picture of a narrow base, constrained by small market size and the high cost of transporting goods. Whilst isolation may cushion islands initially from the effects of wider economic crises, when they are affected the impact may be devastating and the recovery time very long.

Important work done over the past fifteen years by the United Nations Educational, Scientific and Cultural Organization (UNESCO) 'Man and the Biosphere' project on island ecosystems has also stressed the value of an integrated approach. The interactive system formed by man in the environment is viewed by Man and the Biosphere (MAB) researchers as a composition of three linked ecosystems: nature (or the 'support system'), the human population (the 'action system') and culture (the 'regulation system'). Those parts of the action system that integrate environmental resources with spheres of the economy and exchange are termed 'human use systems'; these are the organizations through which resources are managed and in their spatial expression they are rarely if ever congruent with ecosystems. Given the inherent ecological, economic and social fragility of islands, decisions on the development of human use systems should be carried out on the basis of preserving ecosystem resilience. Long-term sustainability becomes more important than what might appear to be economic rationality, since the latter often discounts the welfare and livelihoods of generations to come. Strategies simply to increase production or intensify the utilization of often already overexploited resources – such as building land – may generate substantial earnings initially, but will ultimately lead to depletion and irreversible ecological damage. The cases of desertification following overgrazing or of degradation and pollution of tourist sites are obvious illustrative examples (Vernicos 1987).

Nevertheless, as other chapters in this book will show, planning for island development is beset by extraordinary difficulties. Although the common experience of islands of being 'small, poor and remote' (Selwyn 1978) is not universal, most islands undoubtedly suffer from a range of handicaps – lack of economies of scale, high transport costs, unfavourable terms of trade, difficulty of rational decision-making because of personalism and kinship ties, loss of the educated elite, and so on. Most of these problems are in one sense externally imposed, resulting from islands' poor integration into the modern world economic system. Even the identification of such problems can be considered as an outside imposition. A fact often overlooked is that people living on an island are not automatically concerned with their geographical situation: for them insularity is the norm. Hache (1987:90) points out that the 'island' theme appears hardly at all in the cultural heritage (folklore, songs, poetry) of many islands. When it does appear it is in response to the loss of control over the politics, economy and even language of the islands. Then not only songs but political action may result: demand for autonomy or participation in political networks representing island interests. Both the European Community and the Council of Europe have special committees dealing with islands, whilst at the national scale organizations or federations may play a similar role – the Comhdháil na n-Oileán in Ireland or the Association des Iles du Ponant in France. For academics and planners studying islands, a crucial problem is the failure to develop an appropriate analytical framework for understanding island structure beyond embryonic attempts to operationalize the systems approach. No valid model has been firmly established for the study of islands as a special case within the more generic category of small countries (McElroy *et al.* 1987).

REFERENCES

Alexander, L.M. (1980) 'Centre and periphery: the case of island systems', in J. Gottmann (ed.) *Centre and Periphery: Spatial Variations in Politics*, London: Sage, 135–47.

Beller, W.S. (ed.) (1986) *Proceedings of the Interoceanic Workshop on Sustainable Development and Environmental Management of Small Islands*, Washington: US Department of State.

Boutilier, J.A. (ed.) (1986) *Islands '86: Islands of the World in Perspective*, Victoria, BC: University of Victoria.

Braudel, F. (1972) *The Mediterranean and the Mediterranean World at the time of Philip II*, London: Methuen.

Brunhes, J. (1920) *Human Geography*, London: Harrap.

Buffoni, F. (1987) 'Salina: economic development of an island in the Aeolian archipelago', *Ekistics* 323–4: 158–64.

Carlquist, S. (1974) *Island Biology*, New York: Columbia University Press.

Chapman, R.J.K. (ed.) (1988) *Islands '88: 2nd Conference of Islands of the World*, Hobart: University of Tasmania.

Coccossis, H.N. (1987) 'Planning for islands', *Ekistics* 323–4: 84–7.

Connell, J. (1988) *Sovereignty and Survival: Island Microstates in the Third World*, University of Sydney, Department of Geography, Research Monograph 3.

Dolman, A.J. (1985) 'Paradise lost? The past performance and future prospects of small island developing countries', in E. Dommen and P. Hein (eds) *States, Microstates and Islands*, London: Croom Helm, 40–69.

Dommen, E. and Hein, P. (1985) 'Foreign trade in goods and services: the dominant activity of small island economies', in E. Dommen and P. Hein (eds) *States, Microstates and Islands*, London: Croom Helm, 152–84.

Doumenge, F. (1985a) 'The viability of small intertropical islands', in E. Dommen and P. Hein (eds) *States, Microstates and Islands*, London: Croom Helm, 70–118.

Doumenge, F. (1985b) 'Les îles et micro-états insulaires', *Hérodote* 37–8: 297–337.

Fosberg, F.R. (ed.) (1965) *Man's Place in the Island Ecosystem*, Honolulu: Bishop Museum Press.

Gailey, R.A. (1959) 'Settlement and population in the Aran Islands', *Irish Geography* 4: 65–78.

Gomez Sal, A., Marin, C. and Mendaro, C. (1987) 'Conserving and developing the valuable human landscape of La Gomera, Canary Islands', *Ekistics* 323–4: 170–5.

Gorman, M. (1979) *Island Ecology*, London: Chapman Hall.

Grant, J.F. (1982) *The Lordship of the Isles*, Edinburgh: James Thin.

Hache, J.D. (1987) 'The island question: problems and prospects', *Ekistics* 323–4: 88–92.

King, R. and Strachan, A. (1980) 'The effects of return migration on a Gozitan village', *Human Organization* 39: 175–9.

King R. and Young, S.E. (1979) 'The Aeolian Islands: birth and death of a human landscape', *Erdkunde* 33: 193–204.

Kolodny, E. (1966) 'La population des îles en Méditerranée, *Méditerranée* 7: 3–31.

Kolodny, E. (1974) *La Population des Iles en Grèce*, Aix-en-Provence: Edisud.

Kolodny, E. (1976) 'Aspects d'ensemble de l'insularité méditerranéenne', *Bulletin de l'Association des Géographes Francais* 435–6: 191–5.

Lack, D. (1977) *Island Biology*, Berkeley: University of California Press.

Lockhart, D.G. and Ashton, S.E. (1987) 'Recent trends in Maltese tourism', *Geography* 72: 255–8.

Lowenthal, D. and Comitas, L.(1962) 'Emigration and depopulation: some neglected aspects of population geography', *Geographical Review* 52: 195–210.

MacArthur, R.H. and Wilson, E.O. (1967) *The Theory of Island Biogeography*, Princeton NJ: Princeton University Press.

McElroy, J.L., de Albuquerque, K. and Towle, E.L. (1987) 'Old problems and new directions for planning sustainable development in small islands', *Ekistics* 323-4: 93-100.

Meintel, D. (1983) 'Cape Verde: survival without self-sufficiency', in R. Cohen (ed.) *African Islands and Enclaves*, London: Sage, 145-64.

Miossec, J.M. (1976) 'Croissance et environnement à Jerba', *Bulletin de l'Association des Géographes Francais* 435-6: 203-8.

Murray, D. (1984) 'The nature of island problems', in A. Macartney (ed.) *Islands of Europe*, University of Edinburgh, Unit for the Study of Government in Scotland: 187-92.

Papaioannou, D. (1987) 'Ecological management in the coastal area of Heraklion on the island of Crete', *Ekistics* 323-4: 180-7.

Perpillou, A.V. (1966) *Human Geography*, London: Longman.

Pitt, D. (1980) 'Sociology, islands and boundaries', *World Development* 8: 1051-9.

Pryor, J.H. (1988) *Geography, Technology and War: Studies in the Maritime History of the Mediterranean 649-1571*, Cambridge: Cambridge University Press.

Ramphal, S.S. (1988) 'No island is an island', in R.J.K. Chapman (ed.) *Islands '88: 2nd Conference of Islands of the World*, Hobart: University of Tasmania, 1-28.

Royle, S. (1989) 'A human geography of islands', *Geography* 74: 106-16.

Selwyn, P. (1978) *Small, Poor and Remote: Islands at a Geographical Disadvantage*, University of Sussex, Institute of Development Studies, Discussion Paper 123.

Selwyn, P. (1980) 'Smallness and islandness', *World Development* 8: 945-52.

Semple, E.C. (1911) *Influences of the Geographic Environment*, London: Constable.

Vernicos, N. (1987) 'The study of Mediterranean small islands: emerging theoretical issues', *Ekistics* 323-4: 101-9.

Vidal de la Blache, P. (1926) *Principles of Human Geography*, London: Constable.

Wace, N. (1980) 'Exploitation of the advantages of remoteness and isolation in the economic development of small islands', in R.T. Shand (ed.) *The Island States of the Pacific and Indian Oceans*, Canberra: Australian National University, Development Studies Centre Monograph 23, 87-118.

Wakefield, D. (1977) *Island in the City: the World of Spanish Harlem*, New York: Arno.

Weaver, D.B. (1988) 'The evolution of a plantation tourism landscape on the Caribbean island of Antigua', *Tijdschrift voor Economische en Sociale Geografie* 79: 319-31.

Wilkinson, R.F. (1987) 'Tourism in small island nations: a fragile dependency', *Leisure Studies* 6: 127-44.

Wilstach, P. (1926) *Islands of the Mediterranean*, London: Bles.

3

POLITICAL AND SECURITY ISSUES OF SMALL ISLAND STATES

Anthony Lemon

'What is the right to self-defence without the means of defence? . . . What is the quality of sovereignty if reality dictates the absence of choices?' (Sir Sonny Ramphal, Commonwealth Secretary-General, in his opening address to the first meeting of the Commonwealth Consultative Group on the Special Needs of Small States, 18 July 1984).

The theme of this chapter perhaps requires a word of justification in a book concerned with development. There is, as Sir Sonny Ramphal pointed out elsewhere in the address quoted above, an inextricable link between security and development. Neither national security nor internal political stability are assured in small island states, yet both conditions are important prerequisites of economic development. Equally, economic problems may be at the root of domestic political upheaval, which in turn may encourage external threats to sovereignty and territorial integrity.

The chapter falls into three parts, beginning with a brief survey of the political status of island microstates and of those island territories which are not fully independent, both within and outside the Commonwealth. This is followed by discussion of internal political structures and the relative stability of the governments of Commonwealth small island states. The third section of the paper focuses on the vulnerability of those states to external pressures and threats.

THE POLITICAL STATUS OF ISLAND TERRITORIES

Small island colonies posed a problem in an era of decolonization, especially after the passing by the United Nations (UN) in 1960 of Resolution 1514 which stated, *inter alia*, that 'inadequacy of political, economic, social or educational preparedness should never serve as a pretext for delaying independence'. Prior to 1960 the UN had only six island members: Cuba, Haiti and the Dominican Republic among the original members in 1945; Iceland, rejected by the League of Nations a generation before, was admitted in 1946, primarily in recognition of its strategic value to the United States and its allies in the Second World War (Harden 1985:16); Ireland and Ceylon (now Sri Lanka) followed in 1955. When Cyprus was admitted in 1960 it was the first member with a territory of less than 10,000 square kilometres *and* less than half a million people. In the same year Robinson (1960) observed that it 'was inconceivable that little states could organise in a viable and independent manner if they did not have a population of several million inhabitants'.

In the years which followed the average size of new UN members steadily decreased; the twenty-three states admitted between 1960 and 1983 had an average population of 380,000. Many microstates, especially those in the Caribbean, had tried and found wanting other alternatives to independence, including continuing as a colony with greater self-government, membership of a wider federation and 'associated status'. A few island territories appear satisfied with such arrangements, at least for the time being. Some of the smallest territories remain British dependencies (Table 3.1). The Cook islands pioneered the concept of associated statehood in 1965, and both they and Niue have association agreements with New Zealand. Tokelau voted to separate from Samoa in 1962, and became an overseas territory of New Zealand. It participates in some regional conferences when relevant to its interests, but is within New Zealand's boundaries. These three examples illustrate Dommen's contention that:

Table 3.1 Island states and territories in the Commonwealth

Independent states	Date of independence
Antigua and Barbuda	1981
Bahamas	1973
Bahrain	1971
Barbados	1966
Brunei	1983
Cyprus	1960
Dominica	1978
Fiji	1970
Grenada	1970
Jamaica	1962
Kiribati	1975
Malta	1964
Maldives	1965
Mauritius	1968
Nauru	1968
New Zealand	1907
Papua New Guinea	1975
St Kitts-Nevis	1983
St Lucia	1979
St Vincent	1979
Seychelles	1974
Singapore	1965
Solomon Islands	1978
Sri Lanka	1948
Tonga	1970
Trinidad and Tobago	1962
Tuvalu	1978
United Kingdom	N/A
Vanuatu	1980
Western Samoa	1962

Other island territories	Status
Anguilla	Dependency
British Indian Ocean Territory[1]	Colony
Bermuda	Colony
Cayman Islands	Colony
Channel Islands	Self-governing Crown Possession
Falkland Islands	Colony
Hong Kong	Colony
Isle of Man	Self-governing Crown Possession
Montserrat	Colony
Pitcairn	Colony
St Helena and Dependencies	Colonies
South Georgia and South Sandwich Islands[1]	Colonies
Turks and Caicos Islands	Colony
British Virgin Islands	Colony

Source: Royle (1989):108–9.
Note: 1 Uninhabited except for scientific visits.

> To do without certain legal attributes of independence
> can constitute a practical solution to problems presented
> by geography without affecting the reality of independ-
> ence as it is felt by the inhabitants of the country.
>
> (Dommen 1985:9)

That has not, however, been the general experience in the
Commonwealth Caribbean. Six British Caribbean territories
accepted association with Britain in 1966, with British reserve
powers limited to defence and foreign affairs. But the difficulty of isolating internal from external affairs quickly became
apparent (Thorndike 1987:101), first in Antigua and then most
dramatically in Anguilla, which greatly embarrassed Britain by
its secession from St Kitts-Nevis. It has succeeeded in its wish
to remain a British colony ever since, while the rest of the
'associated states' created in the 1960s were all fully independent by 1981.

Such was Britain's former maritime dominance that all but
ten of the world's forty island states are in the Commonwealth.
A much smaller proportion of non-independent island territories – only twenty-two out of fifty-six – are the responsibility
of member states of the Commonwealth: six are controlled
by Australia, three by New Zealand, and the rest by Britain
(Royle 1989). Four enjoy the status of self-governing regions:
the Faeroes and Greenland in Denmark, the Netherlands
Antilles and Aruba in the Netherlands. Puerto Rico enjoys a
similar position in relation to the United States, though not
full statehood. The US Virgin Islands in the Caribbean and a
string of tiny Pacific islands remain 'unincorporated territories' of the USA. The Federated States of Micronesia, the
Marshall Islands and the northern Mariana Islands are all 'in
free association', whilst Palau is the only part of the former
UN Trust Territory of the Pacific Islands which remains
under the formal jurisdiction of the US Department of the
Interior. Like the USA, France has been noticeably more
reluctant than Britain to grant independence to its island
territories. Guadeloupe, Martinique and Reunion all rank as
full French departments; other French islands are either
overseas territories or 'territorial collectivities', with elected
assemblies having varying degrees of authority. The island
of Mayotte voted in 1976 to remain under French admini-

stration rather than join the Federal Islamic Republic of the Comoros.

Britain's remaining dependencies are, without exception, those which are still seen as too small to become independent, and in most of them there is no significant support for this course. In Montserrat the Chief Minister continued to make statements in favour of independence in 1989, and announced plans for a referendum in 1990, but he later acknowledged that the severe damage caused to the island in September 1989 by hurricane 'Hugo' would increase aid dependence on the UK in the foreseeable future. Only two British dependencies, the Falklands and South Georgia, have been the subject of conflict; Argentina, defeated and forced to withdraw its forces from the islands in 1982, still claims them, although talks in Madrid in October 1989 led to the formal ending of hostilities, the resumption of normal trade, sea and air links, and the reopening of diplomatic relations at consular level. British assets in Argentina have been unfrozen, and Britain has even promised to promote economic links between Argentina and the European Community.

Given the much greater size of some of the island territories still associated with other metropolitan powers, it is not surprising that the status of several is the subject of contention in the islands concerned. The Communist Party of Guadeloupe, which had eight seats out of twenty-six held by the ruling coalition, announced a policy of eventual independence from France in February 1988, although it recognized the material costs implied and would probably not pursue such a policy in practice. In New Caledonia the Melanesian Kanak population strongly desires independence, but this is equally strongly resisted by the French settler population. In June 1988 a tripartite accord was reached between the Kanak Socialist Liberation Front, the (Gaullist) Rally for New Caledonia in the Republic, and the French Government, which provides for a ten-year transitional process leading to a referendum on self-determination in 1998. The accord has been welcomed by the fifteen-member South Pacific Forum.

The main problem among the United States' dependencies is the constitutional future of the Republic of Palau. Currently the US High Commissioner, as Chief Executive, retains only the authority necessary to carry out the obligations and responsibilities of the USA. The problem arises owing to the incompatibility between Palau's anti-nuclear constitution, adopted in 1979, and a compact of free association signed in 1982 as the basis for Palau's self-government, under which the USA would retain responsibility for foreign relations and security, and would have the right to establish military facilities in the territory. The 75 per cent vote which is required to amend the constitution had not been achieved in the seven referenda held between 1983 and 1987, so the compact remained unratified.

SMALL ISLAND POLITICS AND STABILITY

Most Utopian writers, including Plato, Thomas More, Robert Owen and Horace Greeley, visualized their perfect societies as having maximum populations of a few thousand (Hein 1985:17). Climate, beaches and remoteness have given many of the islands under consideration an image of paradise. But small societies, however idyllic their physical environment may seem to the tourist, are not necessarily more harmonious than larger ones, although the nature of political conflict may be different. Schumacher (1973:67) claimed that 'There is no such thing as the viability of states or nations, there is only a problem of the viability of people'. In microstates it is certainly true that the 'viability of people' has a critical influence on socio-economic development and international position.

Sutton (1987) has identified institutional fidelity as a characteristic of small states. The Commonwealth Caribbean states in particular have adopted or adapted the Westminster system, and eight retain the British monarch as Head of State. In several Indian and Pacific Ocean states traditional forms of law and government have been incorporated into constitutions and judicial systems based on Western models. The unicameral legislature of Tonga, for instance, consists of the King, the ten appointed members of the Privy council (which functions as a Cabinet), nine hereditary nobles elected by their peers, and nine popularly-elected representatives. Many island states have sought to accommodate 'out islands', for example, through filling elective or ministerial posts, such as Malta's Ministry for Gozo Affairs, or through local assemblies, such as that of Tobago which has jurisdiction over internal affairs.

Harden (1985) stresses freedom of speech and assembly, respect for human rights and a democratic process enabling the leadership to be changed as political features conducive to stability. She does not see a multiparty system as essential, accepting the validity of a traditional consensual decision-making system backed by sufficient popular participation. Elements of such a system are present in four of the Commonwealth's six 'no-party' island states: Kiribati, Nauru, Tonga and Tuvalu. The Maldives has an assembly of forty-eight members, all but eight of them popularly elected, but no parties, whilst the Sultan wields effective power in Brunei. Elsewhere there is remarkable fidelity to multiparty democracy, which is practised in some two-thirds of the Commonwealth's small island states. Only one, the Seychelles, is a one-party state; its head of state, who seized power in an armed coup in 1979, survived an attempted coup in 1982 and was re-elected for a third five-year term in 1989.

In contrast to the African experience (except in Botswana), multiparty elections in these states have been real not symbolic, enabling peaceful change of government. This has occurred in all island territories of the Commonwealth Caribbean since independence, although Clarke (1987:210–11) attributes the multiplicity of parties to factionalism and the salience of personality in microstate politics. Conflicts among groups are easily translated into personal conflicts, and interests are likely to be either personalized or attached to specific group concerns.

Government and politics are typically pervasive in small island states: everyone tends to depend on the government for something. In so far as this is a function of a high proportion of government employment, Sutton (1987) regards this as a function of the colonial heritage, as it is particularly a feature of Commonwealth states. Such government pervasiveness can lead to partiality in administration and justice, and many states have sought to deal with this in various ways, including constitutional entrenchment of human rights provisions, retention of an appeal to the Judicial Committee of the Privy Council in London (or membership of a common Supreme Court in the Eastern Caribbean), and appointment of an ombudsman (ibid.).

The importance of personality in small island politics has both positive and negative features. Leaders are accessible and likely to be both in touch with their people and held accountable for their actions; voters are in a better position to assess their strengths and weaknesses. But in these small societies it is relatively easy for a determined and ruthless individual to acquire a dominant position, through control of business interests, trade unions, the media and government appointments (Diggines 1985). The regime of Eric Gairy in Grenada is the obvious example, but it is also, fortunately, a relatively isolated one.

Dommen (1980), in an analysis of the year 1975 based on coverage of *The Times* (London), found that islands enjoyed significantly greater internal political stability than continental countries. The experience of small island states since 1975 broadly confirms this view. Successful coups have been carried out in only three – the Seychelles (1977), to which reference has already been made, Grenada (1979 and 1983), and Fiji (May and September 1987).

In Grenada the basic facts are well known: Eric Gairy was overthrown in 1979, and the socialist New Jewel Movement established a People's Revolutionary Government under Maurice Bishop. The latter was murdered in a coup led by hardliners in his own party in 1983, which was followed by a US invasion supported by several Eastern Caribbean states. The wider significance of the invasion will be considered below. The New National Party, a grouping of three centrist parties, won the Grenada election of December 1984, but has been riven by faction-fighting and personality differences since 1987. It retained only one seat in the March 1990 election, in which another centrist party, the National Democratic Congress, won seven seats.

The Fijian coups of 1987 exemplify a factor of internal instability not yet mentioned, namely, ethnic conflict. Fijian Indians, descendants of indentured labourers, marginally outnumber the indigenous Melanesian population, usually referred to as Fijians. However, the latter, represented by the Alliance Party, had managed to control the Fijian Parliament from independence in 1970 until April 1987. The new government was a coalition between the multiracial Fijian Labour Party, and the National Federation Party which was closely allied with Indian interests. The May 1987 coup was staged by the predominantly Melanesian armed forces (with the support of the Fijian chiefs whose influence was threatened by the new government) which established a military regime. Agreement was reached in September to create a racially-bipartisan administration, as a prelude to returning the country to full democracy, but this led to a second coup. Decrees passed during 1988 continued to concentrate effective control of the nation's affairs in Melanesian hands, and to discriminate against those of Indian extraction in the police force, judiciary and civil service. Weapons caches were discovered in June 1988, apparently organized by Indian Fijians in Australia seeking to reverse the May 1987 coup. The country was nominally returned to civilian rule in December 1989, but several army officers were retained.

Fiji's coup also led to the attempted secession of the mainly Polynesian population of the Rotuma group of islands 400 km north of Suva, the capital, who declared independence in January 1988. The Rotumans opposed republican status and were critical of human rights violations by the new government. They wished to become a self-governing protectorate of New Zealand, linked to the British Crown. However, their chiefs were arrested in February 1988 when a small contingent of Fijian troops landed on the islands.

The central conflict in Fiji arises from the existence of a divided plural society rather than from Fiji's character as a small island state. It is true, however, that easy international access and favourable climatic factors encouraged settlement and inward migration in many small islands, leading to both dense populations and ethnic diversity (Commonwealth Consultative Group 1985:16). There are obvious parallels between Fiji and the politics of continental Guyana, where the Indians, who again constitute just over half the population, have been denied power by electoral malpractice and other means. Guyana, in the absence of a violent coup, has remained in the Commonwealth, wheras Fiji has lost its Commonwealth membership and has been told that this and a renewed link with the British Crown cannot be considered until Fiji has an internationally acceptable constitution.

Vanuatu, the former Anglo-French condominium of the New Hebrides, provides another recent example of ethnic and regional conflict in a small island state. This time it was personalized in the form of a struggle between the Prime Minister, representing francophone interests in the capital and Efate atoll, and a nephew of the President who was associated with anglophone interests and the northern outer islands. The President himself attempted a *de facto* coup in December 1988, but was dismissed the following month and found guilty of inciting mutiny among the country's security forces. The personalization of the dispute is characteristic of a small island society, but the basis of ethnic division is unusual in reflecting the dual basis of colonial rule.

Internal strife can come from a variety of causes, most of them by no means confined to small states: ethnic and class divisions, secessionist ambitions, natural disasters leading to economic crises, government inability to deal with economic problems, the divisive effects of migrant labour or political refugees, and major human rights issues. Those small states which are fortunate to avoid all or most of these pitfalls can be much stronger than their size alone would suggest, aided by shared values among their people, firmly-based institutions, and long-recognized borders. A degree of internal cohesion and political stability are important factors aiding the durability of island microstates. States lacking these attributes are more likely to be the target of the external pressures and threats to which we now turn.

EXTERNAL PRESSURES AND THREATS

Small states are notoriously vulnerable to external influence in their affairs, including armed attack, as exemplified by the US invasion of Grenada in 1983, the first military confrontation between a superpower and a small state.

(Clarke 1987:209)

The above statement probably seems uncontroversial to most readers. Yet it uses as an *example* something which has happened for the first time in international affairs. Could it be that the vulnerability of small states is more notorious than it deserves to be? Has too much attention been paid to what might be expected to happen rather than what actually does happen? I am reminded (irrelevantly) of Tatham's (1951) advocacy of the possibilist cause in geography: bananas, he maintained, can be grown in Greenland. Indeed they can, but is it not more pertinent to that old debate that they are not? Island microstates certainly appear vulnerable; but is their sovereignty actually infringed more often or more seriously than that of larger and seemingly less vulnerable states? This is the question I shall seek to consider in the final part of the paper, although without coming to any but the most tentative answer at this stage.

Older works dealing with small states tend to dismiss microstates altogether. Vital (1967:7–8) regards them as 'yet another class of states with reasonably distinct and characteristic problems of their own', whereas Handel (1981:48) more questionably claims that 'methodologically, all the criteria that apply to weak states apply even more readily to mini-states'. Even in relation to the small states which they do consider, both Vital and Handel seem to regard states which survive peacefully as unhelpful to their analyses. Thus Handel (1981:257) observes:

> Weak states can sometimes manipulate and lead a great power against its own will. This has led some observers to exaggerate their power and impact on the international system.

Similarly Vital (1967:190):

> a clear appreciation of small states' condition is made difficult by the fortunate circumstances that the great powers operate today in a climate of thought which promotes caution and hesitation, particularly where the use of force is in question.

Such statements clearly raise the question of what *is* to be regarded as the true condition of small states, what *is* the measure of their power and their impact on the international system?

Vital (1967:184) measures the independence of a state by 'its capacity to withstand opposition and stick to purposes thought commensurate with national interest'. Such 'viability' is, he recognizes, 'a relative quality fluctuating with circumstances, possessed by different states in a different degree, but in no case absolutely and finally as it is by the major powers'. But is it? The United States could not win the war in Vietnam on its own terms, nor could it shape Cuba's foreign policy after Castro took over (Handel 1981:50). Its invasion of Grenada, like Britain's defence of the Falklands, was diplomatically costly: in both cases a different administration might well have been restrained by those costs. Independence is a relative concept, for large and small alike.

Reference to diplomatic costs brings us to the heart of the matter. As the Commonwealth Consultative Group (1985:70) rightly observes,

> It was, after all, the establishment of the United Nations regime, on the basis of equity and co-operation, that originally created the framework for the kind of international community which would be ready to welcome and sustain the presence of very small independent states.

This framework is what Vital (1967:4) refers to as follows:

> The post-war proliferation of small states has occurred in an atmosphere peculiarly conducive to illusions about national strength and to a corresponding reliance on the formal, *legal*, equality of nations.

The principle of sovereign equality of all states is indeed written into the UN constitution, and membership of the UN provides some element of deterrence against predatory neighbours. In the UN, too, a state can alert the international community to a specific security threat. The importance of the UN in this context should not be exaggerated, but it is indisputable that since it came into being there have been few external attempts to take over a small state by force.

What applies to the UN is part of a wider move to what Handel (1981:266) now calls 'new norms' in international relations. These include the right of every nation to self-determination, the sovereign right of states over their own territory, and the increasing disapproval of the use of naked force between states. With nineteenth century colonial exploitation in mind, Handel argues that 'the huge gap that existed between domestically accepted political norms and those governing the conduct of international affairs is slowly closing'. Is it then the behaviour of a past age, pre-dating the existence of island microstates, which has led to the latter being regarded as 'notoriously vulnerable'?

This sounds dangerously like arguing that human behaviour improves as history proceeds. It is in fact arguing that states have collectively come to accept the enduring weakness of human behaviour in terms of international relations, but that they have had a degree of success in collective action to inhibit the application of brute force by increasing the costs of doing so. In this sense the legal equality on which small states are forced to rely is not an illusion, but neither, of course, can it be relied upon to protect them in all circumstances, as the Iraqi invasion and annexure of Kuwait in 1990 demonstrates all too clearly.

Where then does this leave the small island states? They are certainly not without options, but these are likely to be constrained to a greater or lesser degree. If the opposition to a desired course of action is great and the capacity to ovecome it by diplomatic means doubtful, national interest may lie in passivity (Vital 1967). In the same vein the Commonwealth Consultative Group (1985:74) recommended that small states 'should, without appearing to lose any of their sovereignty, adopt a generally discreet posture in the conduct of their foreign policy'. The 'pragmatic conservatism' with which Sutton (1987) characterizes the domestic politics of small states could also be applied to the foreign policy stances of many.

Such policies may be much less restrictive than at first appears. Quite simply, the independence of many, if not most, small island states is not in question at present because they are spared from having to sustain any serious external opposition to their purposes, including the fundamental one of political survival. Policies which appear discreet, passive or pragmatic may be, and often are, those which would be preferred even in the absence of all constraints. One of the landlocked states of southern Africa - islands of another kind - provides a good example. Swaziland's relationship with its powerful South African neighbour has been consensual for most of the time since independence: its relatively good relations with Pretoria have reflected not merely pragmatism on Swaziland's part but also some similarity in ideological world view on the part of a conservative, traditionalist government (Daniel 1984). Despite South Africa's apartheid policies, Swaziland has not sought to resist the implications of dependence, but has in many respects chosen to reinforce them (Lemon 1987).

For Botswana and Lesotho in the same region, dependence on South Africa was more compromising. Economically they have judged it to be in their interests to remain in the Southern African Customs Union (Maasdorp 1982). Politically they have been unable to give to the ANC (African National Congress) the assistance which they would have wished – and have suffered sharp reminders of Pretoria's willingness to use military means when it has suspected them of doing so. But what this means is that Boswana and Lesotho have not been free to undermine the regime of their neighbour. Naturally that has seemed a major constraint because of the nature of the regime in question, but it hardly proves the vulnerability of small states. Even the Lesotho coup of January 1986 would not have occurred had there not been massive internal dissatisfaction with the government of Chief Jonathan: a *de facto* South African blockade of this 'island' within its territory was designed to produce a change of policy not a change of regime, welcome as the latter was to Pretoria. More fundamentally, South Africa has never sought to incorporate Lesotho, despite the years when this uniquely weak neighbour adopted a confrontational stance towards South Africa, rhetorically at least, in the UN and elsewhere. It was protected by its sovereignty, by the reality of Handel's 'changing norms' which made the cost of such action far too high for Pretoria even seriously to consider it.

We can learn still more from the example of these three states, so often described as 'South Africa's hostages'. Swaziland minimized its own regional political role by its choice of a posture so out of tune with the feeling of other states in the region, despite its participation in the Southern African Development Co-ordination Conference (SADCC). Chief Jonathan used a confrontational stance towards South Africa to win support abroad and at home, to minimize the illegitimacy of a government which had rejected the verdict of its electorate in 1970 and held no elections since. Weakened by increasing internal dissent, his government was hardly in a position to play a leading regional role. But Botswana was, and is, very different. With impeccable democratic credentials, a respected leadership and a cohesive society, it has been able to take full advantage of its status as an independent state, notwithstanding its small population and negligible defence

capability. Botswana has maintained an essentially accom-
modationist stance towards South Africa, but

> a liberal ideology coupled with certain practical develop-
> ment possibilities establishes claims on international
> resources. The manipulation of these claims and the
> opportunities created by them in turn strengthens the
> institutions of statehood and makes possible further plays
> of the game.
>
> (Henderson 1974:49)

Botswana's status as a 'frontline state' on the Zimbabwean and
Namibian questions, together with far larger neighbours,
illustrates the potential for a small state to play a significant
international role, at least in the context of its own region.

Much of the literature on small states considers their limited
options in relation to major powers rather than regional
powers such as South Africa. In his second book, Vital (1971)
argues that conflict tends to arise and be pursued precisely
in those 'grey' areas which are not part of a primary power's
recognized domain, but are judged to be peripheral to it.
Conflicts in such regions are likely to result in sharper demarca-
tion of spheres of influence and thus, Vital argues, the decline
of a small state's capacity to call upon one major power for
aid against another: the examples of Grenada and to some extent
Nicaragua are pertinent here. It is therefore in the interests of
the small state to assist 'the retention of regional residual uncer-
tainties' as long as possible. Handel (1981:258) agrees, arguing
the importance for small states of 'a certain condition of tension
and conflict between the powers and an absence of rigidly
defined and mutually neglected spheres of influence'.

Here we may strike a more optimistic note than the Commonwealth Consultative group was able to do in 1985:

> in the last decade the deterioration in international
> co-operation, the widening range of conflict and
> disturbed world economic conditions have noticeably,
> and sometimes critically, aggravated the disadvantaged
> condition of small states.
>
> (Commonwealth Consultative Group 1985:97)

There is no need to labour the extraordinary changes which
have occurred since these words were written. First, the Soviet

Union clearly decided that its involvement in regional theatres of conflict was expensive and largely unproductive. Perestroika within the USSR encouraged changes in Eastern Europe which rapidly acquired an unstoppable and astonishing momentum, of the Soviet Union itself followed by the disintegration into its constituent republics.

These momentous changes must greatly reduce the danger of conflict in Vital's 'grey' areas, and the indirect involvement of both superpowers in regional conflicts. They cannot but represent a substantial improvement in the international milieu from the viewpoint of small states. The Iraqi invasion of Kuwait in 1990 provided the first test of this, and produced the greatest degree of co-operation between the Soviet Union and the United States since 1945. Is it too optimistic to hope that the gloomy prediction of Cohen (1983: 19) that 'perhaps the real crisis, the crisis of increased strategic significance of islands, is yet to come?', has been overtaken by events?

Needless to say, neither Handel's 'changing norms' nor what seems to be the emergence of a new geopolitical order end the vulnerability of small island states. The United States might have less to fear from another Grenada when the perceived threat of Communist expansionism has retreated, but there are other factors which may lead to military intervention in small states, as is clearly demonstrated by the 1990 American invasion of Panama in order to seize its leader, General Noriega, for trial on drug charges. The variety of other threats has been well identified elsewhere (Commonwealth Consultative Group 1985) and need only be summarized here. Military threats to island microstates may come from foreign-based dissident elements, arms or drug traders, or mercenaries. The attempted coup in the Maldives in November 1988 was staged by mercenaries, and foiled with the help of Indian troops, illustrating the value of a well-disposed regional power. Multi-island states are prone to secession, and external forces may be used to foster this. Even without such involvement, secession is often a difficult issue because it raises the problem of what territorial unit has a right to exercise self-determination, as the British know only too well in Northern Ireland.

Non-military threats to territorial integrity are usually non-governmental, but they are much commoner than military threats. They include the use of islands as way-stations in arms and drugs trading (the Bahamas appears to be an instance of the latter), the problems of negotiating with transnational corporations, and the widespread trespass of Exclusive Economic Zones (EEZs) by foreign fishing vessels. Whilst EEZs have brought massive resources within the legal jurisdiction of many island microstates (though others, such as the Leeward and Windward Islands, have lost some of their traditional fishing grounds), such states are ill-equipped to police their new resources. Kiribati, for example, must seek to defend its rights over 3.5 million square kilometres with three small vessels (Dolman 1985:59).

There are no perfect answers to these and other potential threats to small island states. Most investigators of such problems conclude that regional co-operation is potentially the most productive approach, but as Dolman (1985:50) observes: 'In an island region, integration is like spinach: everyone likes it in theory because it is good for you and makes you strong, but few like it in practice'. Such co-operation can be assisted by the resources of the UN and other international bodies, for example, in the gathering of intelligence which can help to anticipate crises at an early stage (Harden 1985: 98–9) and in providing technical support for vetting and negotiating with transnational corporations.

CONCLUSION

A chapter of this length cannot hope to rival the detailed investigations of the Commonwealth Consultative Group and others into the position of small island states. Indeed it has of necessity relied on such sources a good deal for its raw material. What has been argued is that the distinctive characteristics of small island politics need not, and usually do not, compromise either participation or effectiveness in domestic terms, and that these states have a relatively good record of political stability; this is buttressed, in most cases, by a good human rights record and genuine democratic practice. In the wider world it has been seen that the perceived vulnerability of small states derives more from the realities of past than present political conditions. Sovereignty is a major asset in the modern world, especially when strengthened by stability, and many small states stand to gain disproportionately from the easing of global tensions. Small island states may be more constricted than most in the exercise of independence, but their autonomy, and the benefits it confers, appear much enhanced by the possession of formal sovereignty.

REFERENCES

Clarke, C.G. (1987) 'Third World small states: fragile and dependent', *Third World Affairs*, 207-15.

Clarke, C.G. and Payne, T. (eds) (1987) *Politics, Security and Development in Small States*, London: Allen & Unwin.

Cohen, R. (1983) *African Enclaves and Islands*, Beverly Hills: Sage Publications.

Commonwealth Consultative Group (1985) *Vulnerability: Small States in the Global Society*, Commonwealth Secretariat.

Daniel, J. (1984) 'A comparative analysis of Lesotho and Swaziland's relations with South Africa', *Southern African Review Two*, Johannesburg: Raven Press, 228-38.

Diggines, C.E. (1985) 'The problems of small states', *The Round Table* 295: 191-205.

Dolman, A. (1985) 'Paradise lost? The past performance and future prospects of small island developing countries' in E. Dommen and P. Hein (eds) *op. cit.*, 40-69.

Dommen, E. (ed.) (1980) *Islands*, Oxford: Pergamon, (special issue of *World Development* 8 (12)).

Dommen, E. (1985) 'What is a microstate?' in E. Dommen and P. Hein (eds) *op. cit.*

Dommen, E. and Hein, P. (eds) (1985) *States, Microstates and Islands*, London: Croom Helm.

Handel, M. (1981) *Weak States in the International System*, London: Frank Cass.

Harden, S. (ed.) (1985) *Small is Dangerous: Micro States in a Macro World*, London: Pinter.

Hein, P. (1985) 'The study of microstates', in E. Dommen and P. Hein (eds) *op. cit.*, 16-29.

Henderson, W. (1974) 'Independent Botswana: a reappraisal of foreign policy options', *African Affairs* 73: 37-49.

Lemon, A. (1987) 'Swaziland', in C.G. Clarke and T. Payne (eds) *op. cit.*, 156-69.

Maasdorp, G. (1982) 'The Southern African Customs Union – an assessment', *Journal of Contemporary African Studies* 2: 81-112.

Robinson, E.A.G. (1960) *The Economic Consequences of the Size of Nations*, London: Macmillan.

Royle, S.A. (1989) 'A human geography of islands', *Geography* 74: 106-16.

Schumacher, E.F. (1973) *Small is Beautiful*, New York: Harper & Row.

Sutton, P. (1987) 'Political aspects', in C.G. Clarke and T. Payne (eds) *op. cit.*, 3-25.

Tatham, G. (1951) 'Environmentalism and possibilism' in Griffith Taylor (ed.) *Geography in the Twentieth Century*, London: Methuen, 128-62.

Thorndike, T. (1987) 'Antigua and Barbuda', in C.G. Clarke and T. Payne (eds) *op. cit.*, 96-112.

Vital, D. (1967) *The Inequality of States*, Oxford: Clarendon Press.

Vital, D. (1971) *The Survival of Small States: Studies in Small Power/Great Power Conflict*, London: Oxford University Press.

4

GENDER AND ENVIRONMENTAL PERCEPTION IN THE EASTERN CARIBBEAN

Janet Henshall Momsen

To the Europeans who explored the Caribbean in 1492 the islands seemed overwhelming in their attractiveness and munificence. Columbus informed his King that 'these countries far surpass all the rest of the world in beauty'. Today this beauty provides the resource base for the tourist industry which attracts modern visitors to the region. Yet the beauty of the contemporary Caribbean landscape is a chimera. The environment has been severely degraded in the last 500 years and the region is listed as one which has suffered a major loss of plants and animal species (Westermann 1953).

The islands of the Eastern Caribbean are particularly vulnerable because of their small size (Lestrade 1987). Each island is characterized by slightly different flora and fauna and the latter, in particular, tends to be impoverished and imbalanced in comparison with the fauna of the mainland (Watts 1987). Endemic genera and/or species are found on all but the smallest limestone clays but tend to be most numerous on the largest and oldest islands. Small size also accelerates human impact on the landscape and Watts (1966) has shown that in Barbados between 1627 and 1665 virtually all the forest cover was felled and the landscape was transformed.[1]

Elsewhere in the Eastern Caribbean human impact on the environment has been most destructive within the last hundred years. Today forest clearance threatens watersheds in St Lucia and Dominica and derived savanna has become established on Nevis and Montserrat as a result of overgrazing on abandoned estate land. Soil erosion has accelerated, especially on the coralline islands where the formation of new soil from weathering is excessively slow (Hanna 1990). Despite a growing awareness of environmental problems among both the public and policy-makers[2] there is little evidence that the pace of environmental degradation is declining.

This chapter considers recent evidence for gender differences in environmental perception among people living in three zones of Barbados characterized by different environmental threats and also among small farmers in several Eastern Caribbean islands.

ISLAND VULNERABILITY

Small islands are very vulnerable to natural hazards such as volcanic eruptions, as that which befell Martinique in 1902 and St Vincent as recently as 1979, or hurricanes such as the devastating Hurricane Hugo in 1990 which destroyed much of Montserrat and wrought damage estimated at over eight billion US dollars in neighbouring islands. Such hazards can do as much damage to the environment as people, even today, take decades to achieve. Some of the highest population densities per km^2 of agricultural land in the world (740 in Trinidad, 700 in Grenada, 692 in Barbados), despite a migration loss from the region of at least 2.9 million since 1950, have increased the financial costs of these hazards. There is some evidence that global warming is leading to more frequent and severe hurricanes and droughts, two hazards which do seem to be mutually exclusive (Watts 1991). Extreme events such as eruptions or hurricanes may provoke a surge of out-migration and a restructuring of the economy. However, the high profile that a sudden natural catastrophe gives a small island in the world media can attract concentrated inflows of financial and technical aid which may permit improvements in infrastructure that have been needed for years.[3]

The effects of drought and seasonality are also felt more strongly on islands than in continental areas. Food production tends to be characterized by gluts and shortages and, consequently, wildly-fluctuating prices. This lack of price stability discourages farmers from producing for the domestic market. Trade in food products within the Eastern Caribbean, which would help to reduce local price swings, is hindered, not only by high cost and poorly-developed transport facilities, but also by the restrictions necessitated by disease control. However, the possibility of isolating disease together with pests of crops and animals on one island can be seen as beneficial to the region as a whole. Thus the remoteness and isolation of small islands may have both positive and negative effects.

The countries of the Eastern Caribbean are not only faced with problems because they are islands, but their very smallness gives them a very high degree of both economic and ecological vulnerability and limits development options. Small states are characterized by a narrow natural resource base, restricted agricultural potential, small domestic markets and limited employment opportunities which effect human resource utilization by provoking both a brain drain and high un-employment rates. Dependence on a limited range of exports enhances the vulnerability of small states to fluctuations in world demand and to price conditions which are exacerbated by failures to fill quotas because of environmental extremes. Debt problems make it difficult for island governments to sacrifice traditional export-oriented agriculture for the development of better, more sustainable agricultural systems such as small-scale polyculture (Kimber 1991). In addition, the limited infrastructure and lack of both a skilled work-force and an entrepreneurial class in many Eastern Caribbean countries, has reduced the area's ability to take advantage of concessionary trading agreements under the Lomé Convention, the Caribbean Basin Initiative and the CARIBCAN agreement (Momsen 1992). As Demas said two decades ago 'the alternatives open to small countries in the contemporary world are much more narrowly circumscribed than those open to larger countries' (Demas 1965:39).

Development in these countries depends on an understanding of the environment. In order to examine aspects of gender and the interaction of the environment and development, this chapter focuses on recent changes in agricultural land use and on environmental perception in the Eastern Caribbean.

LAND USE CHANGE

The structure of agriculture in the Eastern Caribbean remains largely as it was in colonial times with export agriculture and large holdings predominant (Table 4.1). Beckford (1975) saw this alienation of a nation's resources from its people as a prime cause of Caribbean underdevelopment. However, a new problem is the loss of farmland to non-agricultural uses. In Barbados the area of farmland declined from 33,000 hectares to 19,000 between 1950 and 1986 (Cumberbatch 1987); in St Lucia it fell from 35,000 hectares in 1961 to 28,000 in 1973, and in Grenada from 28,000 hectares in 1946 to 13,000 in 1981. Some of the farmland has been lost to other uses such as housing, roads, airports and tourist-related activities, but much has been abandoned because of environmental degradation, such as soil erosion and weed infestation. Other land is being used inappropriately resulting in land slips, reduced soil fertility, irregular spring flow and lower rainfall (Corker 1986). Not only has there been a decline in agricultural land, but there has also been a decline in the proportion of such land actually cultivated, especially on large estates. Sugar-cane cultivation has decreased throughout the region and has generally been accompanied by a fall in per capita agricultural output (Hagelberg 1985).

Table 4.1 Peasants and plantations in the Eastern Caribbean

	Small peasant farms with under 2 hectares of land		Plantations with over 200 hectares of land	
	% of farms	% of farmland	% of farms	% of area
Barbados (1971)	98.3	11.6	0.4	43.0
Dominica (1972)	73.0	12.6	0.3	30.6
St Lucia (1973)	82.1	14.2	0.1	47.4
St Vincent (1972/3)	87.0	23.5	0.1	40.4
Grenada (1981)	88.3	30.1	0.1	15.0
Antigua (1973/4)	91.1	26.7	0.3	42.2
Montserrat (1972)	90.0	19.8	0.2	34.9
St Kitts–Nevis (1975)	95.2	40.0	0.1	6.5

Source: Agricultural censuses of the respective countries.

One of the reasons for the decline in agricultural production was the feminization of agriculture which occurred throughout the region in the 1960s largely as a result of male migration overseas. The proportion of female farmers on small farms at this time ranged from 56 per cent in Montserrat, and 53 per cent in Barbados to 43 per cent in St Lucia, and 36 per cent in Martinique. The feminization of agriculture without the necessary institutional support led to decreased intensity of production and land left idle because of labour shortages. Chaney (1983) has suggested that it also led to many of the region's nutrition problems being exacerbated by recent economic restructuring which has put increasing pressure on women (Deere *et al.* 1990). Women farmers are more likely than men to receive remittances from relatives working overseas (Momsen 1986) and analysis of data from surveys in Montserrat and Nevis showed a strong correlation between a high dependence on remittances and under-utilization of land.

WOMEN'S ECONOMIC ROLES IN THE CARIBBEAN

Women in the Caribbean exhibit higher levels of labour force participation than in most parts of the Third World; indeed, these levels are often above those of many industrialized nations (Table 4.2). Economic activity rates are highest in the French islands of Martinique and Guadeloupe and lowest in the Hispanic Caribbean. Within the Anglophone Caribbean, proportions are lower in Trinidad and Guyana where there are many people of Indo-Caribbean heritage.

These regional differences reflect a tradition of economic autonomy among Afro-Caribbean women which may be attributed to three historical factors. First, under slavery, women were expected to perform much of the heaviest field labour and then, after emancipation, male out-migration left many women alone to carry the burden of supporting their children (Momsen 1988). The restrictions on marriage enforced by slave-owners weakened the conjugal ties while often leaving the mother-child bond intact, although some research suggests that these effects may not have been as widespread as previously thought (Higman 1984). Sex-specified migration further encouraged the development of female-headed households. In the Eastern Caribbean, on average, 35 per cent of households are headed by women. There is considerable regional variation ranging from half of all households in St Kitts-Nevis, 44 per cent in Barbados and 42 per cent in

Table 4.2 Caribbean statistics with special reference to women

State	Area (kms²)	GDP/ capita (US$ 1987)	Population 1989 ('000)	% population female	% economically active population (female)	Caloric daily capita intake	Fertility rate
Anguilla	91	2,900	7.3	50.9	40.5	N/A	1.9
Antigua & Barbuda	442	2,570	78.4	52.00	40.1	2,089	1.7
Aruba	193	6,810	61.3	51.62	36.7	2,925	1.8
Bahamas	13,939	10,320	249.0	50.20	44.5	2,699	2.6
Barbados	430	5,330	255.0	52.15	47.2	3,181	1.8
Belize	22,965	1,250	18.5	49.34	32.5	2,585	5.4
Bermuda	54	23,100	58.8	51.28	46.8	2,545	1.8
British Virgin Is.	153	9,490	12.5	49.9	38.8	N/A	2.2
Cayman Is.	264	17,390	25.3	51.4	43.8	N/A	1.6
Cuba	110,861	2,690	10,540.0	49.65	35.8	3,107	1.8
Dominica	750	1,440	82.8	50.19	34.1	2,649	3.1
Dominican Republica	48,443	730	7,012.4	49.18	28.9	2,464	3.8
French Guiana	86,504	2,130	95.0	47.34	45.5	2,747	3.1
Grenada	345	1,340	96.6	51.80	N/A	2,409	3.1
Guadeloupe	1,780	3,490*	341.0	51.20	46.6	2,674	2.2
Guyana	215,083	380	754.0	49.84	29.9	2,456	2.8
Haiti	27,400	360	5,520.0	51.52	40.9	1,902	4.6

Jamaica	10,991	960	2,376.0	50.25	45.3	2,581	2.9
Martinique	1,091	4,280*	337.0	51.52	50.6	2,780	2.1
Montserrat	102	2,530*	12.0	51.90	42.7	N/A	1.9
Netherlands Antilles	800	6,810*	183.0	51.32	41.8	2,925	3.4
Nevis	93	1,700	9.7	49.66	41.0	2,349	3.0
Puerto Rico	9,104	5,520	3,308.0	51.32	37.5	N/A	2.4
St Kitts	176	1,700	34.4	49.66	41.0	2,349	3.0
St Lucia	617	1,370	150.0	51.46	39.1	2,776	3.8
St Vincent	389	1,070	114.0	51.56	36.1	2,499	3.0
Surinam	163,820	2,360	405.0	50.40	27.2	2,713	3.0
Trinidad & Tobago	5,128	4,220	1,285.0	49.95	33.8	3,058	2.8
Turks and Caicos Is.	500	4,780	13.5	51.7	42.8	N/A	3.8
US Virgin Islands	352	10,050	107.0	52.15	45.5	N/A	2.6
United Kingdom	244,110	10,430	57,218.0	51.23	41.4	3,218	1.8
United States	9,529,063	19,860	248,777.0	51.29	44.3	3,642	1.8
USSR	22,403,000	8,160	287,800.0	52.74	49.9	3,205	2.4

Source: Encyclopaedia Britannica (1990); *Book of the Year*, Chicago.
Note: * 1985 data.

Montserrat, to one-quarter in Trinidad and Tobago where the lower figure reflects the influence of the Indo-Caribbean conjugal family (Buvinic and Youssef 1978:99; Caricom 1980–1).

In many cases female-headed households were able to survive economically because they had access to land for subsistence production. In the Anglophone Caribbean women inherited land in their own right and were also often given usufruct land belonging to migrant male relatives. More recently women have seen education as their main source of security. In Montserrat, a higher proportion of women than men, under 34 in 1980, had completed secondary education (Caricom 1980–1). In Barbados in 1980 14.5 per cent of adult women but only 12.2 per cent of men, as compared to 6.3 per cent of women and 7.0 per cent of men in 1970, had passed some secondary school leaving examinations. It is also noticeable that although there are more adult men than women with university degrees in the Barbadian population, for the cohort aged 15 to 24 in 1980, there are more women than men with a university education (Caricom 1980–1). In many cases where men have sought economic opportunity through migration, women have achieved it through education.

On these small islands the distinctiveness of rural and urban areas is much less than in most parts of the world. Farm families are often involved in urban employment even while they continue to cultivate the land. Women on these farms may have several roles in addition to their domestic duties: they may be the decision-makers on the farm or they may assist the farm operator by providing labour, they may buy and sell farm produce locally or on a neighbouring island, or they may work in a non-farm job. Individuals may fulfil all these roles during their lifetime but they are more likely to do so sequentially than simultaneously. Occupational multiplicity is far less common among women farmers than among men.

Within these broad patterns life stage is influential. Fewer women than men have off-farm jobs and they are most likely to be involved in this type of work between the ages of 35 and 49. In St Vincent and Nevis, two of the least developed islands, lower percentages of women than men hold skilled off-farm jobs, and farming remains important for them as a source of income and to provide food for the family.

Particularly in St Vincent, the poorest island, older women farmers are very dependent on supplementary income from working as petty traders. In more developed Barbados there is little gender difference in skill levels of off-farm employment, and on the farm we see an increase in 'hobby farming' which allows those with full-time off-farm jobs to enjoy their own fresh produce.

GENDER PERCEPTIONS OF THE ENVIRONMENT

In order to understand the public's responses to national policy initiatives it is necessary to appreciate perceptions of the environment. This section is based on a survey of 175 Barbadians living in urban, rural and coastal tourist settings in 1989. Comparisons are made with earlier field surveys of the perceptions of small farmers in the Leeward Islands of Montserrat and Nevis and the Windward Island of St Lucia.

As has been shown, Barbados is a densely-populated island with a highly-educated population and considerable public awareness of environmental problems. Soil erosion in the Scotland District, where the coralline cap has been worn away exposing the easily eroded underlying clays, is highly visible resulting in land slips, house collapse and road washouts. For three decades this problem area has been in the hands of a government agency with powers to control land use and to introduce improvement measures. It was assumed that awareness of soil erosion as a problem would be higher among people living in the Scotland District than among those elsewhere where soil erosion takes the form of sheet or gully erosion and is less immediately noticeable. Along the coast fish stocks are being depleted, the sea is becoming polluted and the coral reef is deteriorating. On the west coast tourism development has led to a high density of construction of new facilities.

It was hypothesized that there would be regional differences in the perception of different environmental problems and that gender differences would also be apparent. The survey included 85 men and 90 women respondents with average ages of 39 years (men) and 41 years (women). When asked if they thought soil erosion was a serious problem in Barbados 61 per cent responded positively. However, gender differences were very noticeable with 74 per cent of men, but only 48 per cent of women, agreeing. Of the 49 respondents living in the Scotland District, where erosion is so conspicuous, almost the same proportion, 62 per cent felt that it was a problem. Yet here gender differences were even more marked with 82 per cent of the men, but only 45 per cent of the women respondents, saying that they thought it was a serious problem. Among the 15 per cent of the people surveyed who were small farmers, the gender difference was less marked and a higher proportion of both men (83 per cent) and women (66 per cent) declared that erosion was a major problem. For both women and men, just under half the farmers were aware of anti-erosion measures they could take to protect their land.

The main source of renewal for Barbadian soils is ash from volcanic eruptions on neighbouring islands. The most recent ashfall came from an eruption on St Vincent in 1979. Of the people surveyed, 57 per cent of the women and 62 per cent of the men remembered the event but only 31 per cent of the women and 51 per cent of the men thought that the ash was important for soil fertility.

Although Barbados has not experienced a serious hurricane since Hurricane Janet in 1955, most Barbadians are very aware of the danger from such events and can name several recent hurricanes in the region. However, in the survey more men (18 per cent) than women (9 per cent) felt that the last hurricane had caused major damage on Barbados.

Awareness of sea pollution is increasing as local fish become less easily available on the island and oil slicks and garbage can be seen close in shore. However, many rural Barbadians, especially older people, rarely go to the beach and so felt themselves unable to comment about such pollution. Some 56 per cent of men and 48 per cent of women respondents felt that the sea was becoming polluted, with the fishermen in the survey being especially knowledgeable. Many also felt that the beach was becoming too crowded but others felt that the tourist hotels kept the beaches very clean.

It might have been expected that the impact of tourism on the environment would be resented, but on the whole both men and women were very positive about the built environment of the tourist zone. About half of the Barbadians surveyed felt that tourism had led to both increased prices and higher levels of employment and wages. Clearly government policy stressing the benefits of tourism to the island has been successful.

Generally gender differences in environmental perception were most marked among the older, less-educated survey respondents. Older women were most likely to be fatalistic and to depend on the efficacy of prayer rather than technical measures to overcome environmental problems. Women also tended to feel that traffic fumes and garbage collection were more important environmental problems than soil erosion, reflecting the different social environments of men and women. Where both men and women were involved in farming, gender differences in perception of erosion were much less marked.

Yet the problems faced by women and men small farmers do differ. Table 4.3 compares field data collected in the 1970s in Montserrat, Nevis and St Lucia and differentiates between

Table 4.3 Gender differences in perception of problems by small farmers (percentage of farmers mentioning problem)

Problem	Montserrat (1973)		St Lucia (1971)		Nevis (1979)	
	Male (n = 44)	Female (n = 22)	Male (n = 47)	Female (n = 10)	Male (n = 70)	Female (n = 30)
Availability of labour	27	64	38	80	47	57
Transport and roads	23	18	36	60	7	17
Markets	9	9	58	50	80	83
Pests and disease	55	73	68	60	63	73
Livestock damage to crops	5	14	NA	NA	16	3
Livestock killed by dogs	NA	NA	NA	NA	44	27
Theft	3	0	NA	NA	14	27
Credit availability	16	0	22	0	34	40
Availability of agricultural inputs	3	5	0	0	16	17
Water shortage	41	36	38	40	17	23
Land availability	23	18	12	0	10	10
Soil erosion	16	14	38	50	13	13
Stony soil	11	5	0	0	1	0
Land tenure	5	0	6	10	1	0
Physical infirmities	5	14	6	0	4	0
No problems	9	0	2	0	NA	NA

Source: Field-work

the ranking given to problems by men and women farmers. In general it appears that women farmers experience greater difficulties in obtaining labour than men because male workers are often unwilling to work for a female farmer and women farmers are less likely than men to be able to pay for hired labour. In all other aspects of farming, differences between the islands were more important than gender differences in problem perception. Environmental differences between the islands are revealed by variation in the relative importance of soil variables. There was little gender difference in awareness of soil erosion problems, although more men than women seemed to feel that stony soil was a problem. Pests and diseases were of major importance for both men and women farmers on all three islands.

CONCLUSION

Gender differences in environmental perception appear to be related to the social and economic roles of men and women. Both are aware of major environmental disasters such as hurricanes and eruptions. More insidious problems such as soil erosion and marine pollution are most strongly perceived by those dependent on these resources such as farmers and fishermen. Women were more aware of problems which affected their daily domestic routine such as garbage collection and traffic pollution. There seemed to be limited perception of the need to protect the sea and sand resources which are the basis of the tourist trade. Attempts to move toward sustainable development based on careful use of the renewable resources of the land and sea depend on greater public awareness of environmental problems.

The peoples of the West Indies have not really adjusted to the harsh demands of 'an apparently beneficent but in reality a difficult and in many ways an unpredictable environment' (Watts 1987:338). Once environmental deterioration is established it appears to be little affected by changes in economic activities but accelerates under its own momentum. The challenge for the region is to widen the perception of both men and women of the long-term effects of all types of environmental degradation.

NOTES

1 A specimen of the mastick (*Mastichodendron sloaneanum*), a prized timber tree in seventeenth century Barbados which had long been thought to have disappeared, has been identified in 1991 growing in a gully in the northern part of the island (J.C. Hudson, personal communication, April 1991).
2 The Caribbean Development Bank organized a series of lectures on the region's environment in 1989 and the recently-founded Barbados National Trust is awakening public interest in the island's natural and historic landscape.
3 During the 1980s Dominica received financial and technical assistance from many countries triggered by the devastation caused to the island at the beginning of the decade by two hurricanes. Montserrat has been able to upgrade its housing standards because of free housing given to replace dwellings destroyed by Hurricane Hugo.

REFERENCES

Beckford, G.L. (ed.) (1975) *The Caribbean Economy*, Mona, Jamaica: Institute for Social and Economic Research, University of the West Indies.

Buvinic, M. and Youssef, N.H. (1978) *Women-headed Households: The Ignored Factor in Development Planning*, Washington DC: International Centre for Research on Women.

Caricom (1980-1) *Population Census of the Commonwealth Caribbean* (Montserrat, Barbados), Kingston, Jamaica: Statistical Institute of Jamaica.

Chaney, E.M. (1983) 'Scenario of hunger in the Caribbean: migration, decline of smallholder agriculture and the feminization of farming', *Working Paper no. 18*, East Lansing: Michigan State University.

Corker, I.R. (1986) *Montserrat, A Resource Assessment*, Surbiton: Overseas Development Administration, Land Resource Development Centre.

Cumberbatch, E.R. St J. (1987) 'The land - our most precious resource', *The Bajan*, August-September: 12-19.

Deere, C., Antrobus, P., Bolles, L., Melendez, P.P., Rivera, M. and Safa, H. (1990) *In the Shadows of the Sun: Caribbean Alternatives and U.S. Policy*, Boulder, Colorado: Westview Press.

Demas, W.G. (1965) *The Economics of Development in Small Countries with Special Reference to the Caribbean*, Montreal: McGill University Press.

Hagelberg, G.B. (1985) 'Sugar in the Caribbean: turning sunshine into money', in S.W. Mintz and S. Price (eds), *Caribbean Contours*, Baltimore: Johns Hopkins University Press, 85-126.

Hanna, L.W. (1990) *Land Use Changes in the Island of Barbados 1985-90*, Seminar Paper no. 61, Department of Geography, University of Newcastle upon Tyne.

Higman, B.W. (1984) *Slave Populations of the British Caribbean, 1807-34*, Cambridge: Cambridge University Press.

Kimber, C. (1991) *Changing Caribbean Ecosystems*, paper presented at the Conference on Alternatives for the 1990s Caribbean, London, January 1991.

Lestrade, S. (1987) 'Economic issues affecting the development of small island states: the case of the Caribbean', *The Courier* 104: 78-84.

Momsen, J.H. (1986) 'Migration and rural development in the Caribbean', *Tijdschrift voor Economische en Sociale Geografie.* 77: 50-8.

Momsen, J.H. (1988) 'Gender roles in Caribbean agricultural labour', in M. Gross and G. Heuman (eds) *Labour in the Caribbean*, London: Macmillan, 141-58.

Momsen, J.H. (1992) 'Canada–Caribbean relations: wherein the special relationship?', *Political Geography*, 11(5): 501-13.

Watts, D. (1966) *Man's Influence on the Vegetation of Barbados 1627 to 1800*, Occasional Papers in Geography no. 4, University of Hull.

Watts, D. (1987) *The West Indies: Patterns of Development, Culture and Environmental Change since 1492*, Cambridge: Cambridge University Press.

Watts, D. (1991) *In the Nature of Disasters: Implications for Caribbean Development*, paper presented at the Conference on Alternatives for the 1990s Caribbean, London, January 1991.

Westermann, J.H. (1953) *Nature Preservation in the Caribbean*, Foundation for Scientific Research in Surinam and the Netherlands Antilles, Utrecht.

5

TOURISM DEVELOPMENT IN SMALL ISLANDS

Past influences and future directions

Richard W. Butler

INTRODUCTION

Small islands, for a variety of reasons, have long been viewed as attractive destinations for both recreational and touristic purposes. Their appeal may relate to the very real feeling of separateness and difference, caused in part by their being physically separate, and perhaps therefore different, from adjoining mainlands. Where such physical separateness is accompanied by political separateness, the appeal can be expected to increase, and given people's desires for the different while in pursuit of leisure, different climates, physical environments and culture can all be expected to further the attractiveness of islands as tourist destinations.

Small islands in particular would seem to be even more attractive to tourists. The reasons for this can only be speculated, and may range from the preference to visit places which one can 'get to know' in the week or two weeks one is there as a tourist, to the feeling that small is likely to be more 'authentic', less developed or commercial than large, and even to what we might call the 'Robinson Crusoe' factor, the tourist being one of only a few people to experience the attractions of the location. Many brochures and much tourist information feature this latter idea, access to 'one's own deserted tropical island' to quote a recent Garuda airline advertisement. Irrespective of the exact nature of the appeal, psychological, geographical, economic, climatic, sensual or even convenience, many islands, especially small islands, have become international tourist destinations. A disproportionately large number qualify as 'tourist countries' under Bryden's (1973) definition, and nowhere is this more true than in the Caribbean, both in absolute and in relative terms. Almost all of the Caribbean island states qualify under both criteria, and tourist visitation in 1987 totalled over 9.6 million.

Not surprisingly, the development of tourism on islands has attracted a reasonable amount of interest in the academic community in recent years, heightened inevitably by the rapid increase in the level of, and the impacts resulting from, tourist development and visitor arrivals. As Pearce (1987) points out, however, much of the research which has been done on island tourism has tended to be of the case-study variety, with a somewhat marked gap in research of an overall conceptual nature.

In an almost unique review of some general principles and patterns of tourism development on islands, Pearce notes some common elements in a variety of areas. His discussion includes the frequent clustering of accommodation close to the primary urban centre and/or the airport, and in coastal areas of high amenity, and the development of enclaves, both physically and in economic terms. Pearce also notes the constraints which tend to be a result of smallness, including limited resources, a small population and domestic market, diseconomies of scale, and a reliance upon foreign trade in a limited number of products (Pearce 1987:154). In the context of tourism, the major appeal is normally a combination of sun, sand and sea, plus possibly sex and shopping. The nature of the tourism product, plus many common problems of access, especially on volcanic islands, tends to emphasize the coastal nature of the tourism development, often linked by, and dependent upon, a coastal circular road. Only in larger islands, or those with marked relief and hence climatic variation, is there much interior development, for example, of hill stations to avoid the extreme summer heat, for example, in Jamaica and Cyprus or to allow alternative forms of tourism such as skiing in Cyprus or Hawaii.

The lack of a large local population in many cases, with resulting lack of development of internal domestic tourism,

Cuba being one exception to this, has meant a general lack of expertise and infrastructure to accommodate the international tourist. The lack of public services in many small islands also serves to concentrate development where and when it does take place (ibid.:157). Finally Pearce notes that most island tourism is now dependent upon air transportation for access, which to a degree limits the types of visitor likely to come, and makes it relatively easy and attractive for vertical integration of the development, and the appearance of the package tour and tourist. Such developments again tend to encourage enclave and concentrated development (ibid.:160).

In recent years more general reviews of tourism development on islands have appeared, with attention being focused particularly upon the Caribbean (for example, Edwards 1988; Hills and Lundgren 1977; Holder 1979; Deward and Spinard 1982) and upon the South Pacific (for example, Britton and Clarke 1987; Pearce 1980 and Rajotte 1982). While the Mediterranean islands may have a longer history of use as tourist places, emphasis in research has still tended to focus upon them as individual or small group entities. Some indication of the importance now attached by researchers to tourism in islands can be judged by the fact that Wilkinson (1988) was able to produce a bibliography of over six hundred references pertaining directly and indirectly to tourism in island microstates. Despite this evidence of the amount of research related to tourism on islands, one might note the overall general reluctance of many researchers to acknowledge the real significance and level of importance of tourism to many island states and Third World countries in general. In the volume *The Geography of the Third World* (Pacione 1988), for example, there is no chapter dealing with tourism, and in Connell's chapter on 'Contemporary issues in island microstates', only two of thirty-five pages are devoted to tourism, despite the author noting that tourism is 'much the largest contributor to the G.D.P.' in four countries cited, and 'the massive dependence upon it' of many island microstates (Connell 1988:448).

Related to this background there are other factors, which, it is suggested have played significant roles in the nature and extent of the development of tourism in many island states. These are discussed below in relation to a general model of tourist development articulated by the author (Butler 1980).

CONCEPTUAL MODEL

The assumption behind the model illustrated in Figure 5.1 is that tourist destination areas are essentially organic, that is, they follow a similar process of growth and change over time, irrespective of their spatial location or, to a large extent, the basis of their attractivity. This is not to say that their development is not, or cannot be altered or capable of alteration by man, but that a general process or cycle of evolution can be identified. The asymptotic or S curve depicted in Figure 5.1 is clearly analogous to the product life cycle utilized in the development and marketing of many products. The analogy is felt to be a realistic one since it can be argued strongly that tourist destinations are products in the very real sense of the term, and are developed, marketed and sold as most products are. The key differences are that the product must be 'consumed' *in situ* by the customer, although visual and other souvenirs/reminders may be taken home, and the pattern of consumption is more akin to rental than outright purchase. An analogy can also be drawn to the pattern of growth and decline of wildlife populations under natural conditions in the absence of predators. The population grows slowly, then exponentially, and finally crashes as it overtaxes food supply and perhaps becomes subject to disease.

The development of the model and its predecessors in the literature have been discussed elsewhere (Butler 1980) and only a brief discussion will be made here. Five consecutive stages are identified, with possible variations for the sixth stage. These stages or periods of development are not meant to be completely mutually exclusive but rather descriptive of the general nature of the development cycle at a point in time.

It is suggested that the general pattern of development is characterized by a period of *exploration*, in which small numbers of visitors 'discover' a destination, making individual, non-institutionalized travel arrangements, utilizing facilities provided for, used and owned by locals. This stage

74

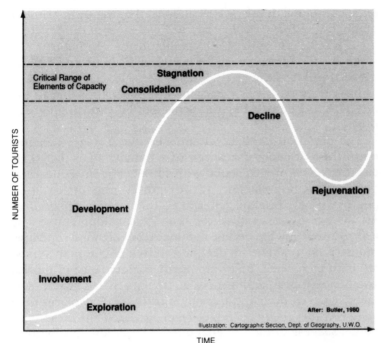

Critical Range of
Elements of Capacity

Stagnation

Consolidation

Decline

NUMBER OF TOURISTS

Rejuvenation

Development

Involvement

Exploration

After: Butler, 1980

Illustration: Cartographic Section, Dept. of Geography, U.W.O.

TIME

Figure 5.1 Tourist area cycle of evolution

is followed by one of *involvement*, characterized by greater, and perhaps regular visitation, and in which local entrepreneurs begin to provide new facilities or convert existing facilities for visitors.

The most significant and obvious changes begin to take place in the *development* or 'take-off' stage, which sees the establishment of much of the infrastructure (services, accommodation and facilities) for tourism, the appearance of defined and regular markets, stimulated by advertising, and the general replacement of smaller locally-controlled facilities by larger non-locally-owned development. The *consolidation* phase represents the last stage of initial growth, albeit at a rate below that during the development phase, with control of much or all marketing, development and visitor services being located in the origin-areas of visitors or elsewhere in the global economic 'core'. It is in this period that the absolute maximum number of visitors may be recorded and most of the development will be 'secondary' or complementary in nature. The *stagnation* stage represents a period of little or no absolute growth, and capacity levels of many variables can be expected to have been reached or exceeded. Problems relating to the physical and human environments can be expected by this stage.

The direction of the cycle in subsequent stages depends upon the presence or absence of a number of variables. If capacity levels are exceeded with subsequent environmental degradation, a reduction in attractivity and hence competitiveness can be expected, and visitor numbers will *decline*. An absence of new capital input and redevelopment will be a likely result and the decline continues. In a highly competitive industry such as tourism decline and lack of competitiveness are hard to reverse. Redevelopment, maintenance of quality and hence appeal, and respect of capacity levels, that is, sustainable development, can result in *stability*. Few places have achieved this over the longer term and even for these, perpetual success cannot be guaranteed. For a few destinations *rejuvenation* may occur, if sufficient capital and redevelopment occurs the image and attractiveness is either restored or changed dramatically and successfully marketed. The addition of a powerful new attraction, such as gambling in Atlantic City, or the development of a new holiday season, as in skiing at Aviemore in Scotland, may be a sufficient catalyst, although problems, especially those related to saturation and overuse will not disappear (Stansfield 1978). Evidence from a number of locations (Meyr-Arendt 1985; Sadler 1988; Weaver 1988; Wilkinson 1987) would suggest that as a general model the cycle described above has considerable validity.

Table 5.1 summarizes some of the elements characteristic of the stages of the cycle and also integrates the types of tourists which could be expected to be present at the various stages in the cycle, drawing on the work of Cohen (1972) and Plog (1972). It should be acknowledged that the model best reflects destination areas which have developed from existing communities, and that certain specific destinations, for example, Cancun in Mexico, represent 'instant' resorts which have no exploration or involvement phases. Even for these locations,

Table 5.1 Hypothetical cycle of tourist areas

Stages	Tourist numbers	Facilities	Contact	Change	Control	Tourist type (Plog 1972)	Tourist type (Cohen 1972)
Exploration	Very small	Few	Low	None	Local	Allocentric	Drifter
Involvement		Local			+ Regional	Near-allocentric	
Development	Maximum	Non-local	High Impersonal	Major	National/ international	Mid-centric	Individual Mass
Consolidation		Peak			All Levels	Near-psychocentric	
Stagnation				Stable		Psychocentric	Organized
Decline	Small	Local	High Mechanical	Relics	Increasing Local	Mid-centric	Mass
Rejuvenation		New		Major	National/ international		Individual Mass
Conversion	High		Low				

it is suggested that once established they will be subject to the same pressures and experience the same subsequent cycle of development. What is important is the possible inevitability of progression, raising the question asked by Plog (1972) as to whether destination areas carry with them the seeds of their own destruction.

In the context of the Caribbean one may briefly suggest that destinations such as the Bahamas and Puerto Rico represent locations in the consolidation or close to the stagnation phase, those such as Barbados and Jamaica in consolidation, the British Virgin Islands and Aruba in development, St Vincent in involvement, and Anguilla in exploration. Destinations such as Grenada, Haiti, Jamaica and Cuba represent individual and interesting variations reflecting somewhat unique historical variations and circumstances.

Cuba, until the Communist revolution, had progressed further through the cycle than any Caribbean destination and was by that time already in late development, with considerable evidence of environmental and social corruption and degradation. The Communist revolution and subsequent prohibition of travel by US residents to Cuba served to decimate the tourism industry, its survival in a drastically-reduced form only made possible by flows of Eastern Europeans. Developments in the 1970s and 1980s have seen a rejuvenation of tourism in Cuba, which could change again if political circumstances alter (Hinch 1990). Grenada, similarly, experienced the negative effects of ideology upon tourism, aggravated by violence, political unrest and invasion. Tourism in Grenada is recovering but the island has not yet overcome the image problems of the mid-1980s.

Haiti has been plagued by potential violence and unrest since its creation by revolution. Tourist development has tended to be in enclaves as separated from potential civil unrest as possible, but its pattern of development has inevitably been spasmodic. The political events of 1989 may represent the beginnings of stability but a considerable time-lag is likely before confidence is complete and tourist numbers rise regularly. Jamaica to a degree represents the Caribbean in miniature. The oldest developments around Kingston and some 'great houses' of the past have for the most part declined and disappeared. Resort areas of Ochos Rios and Montego Bay have proceeded through their own cycle and are now in consolidation or even stagnation, facing problems of relatively-declining appeal compared to newer developments. Other, smaller communities in Jamaica are still in the development stage, while other parts are still in the exploration phase. Given the presence of modern air links and infrastructure and Jamaica's reputation and image in the tourism world, it may be unlikely that other areas will proceed through the early stages of the cycle, but may go straight to development. For more detailed discussion of the development process, three island states are examined in more detail.

CASE STUDIES

Mass tourism, as the modern phenomenon is characterized, is primarily a post-Second World War feature in the Caribbean. However, the process of visitation to the Caribbean for pleasure in a variety of guises began much earlier, coinciding with the flourishing plantation system of agriculture. The Caribbean, the first part of the 'New World' to be discovered, was deliberately depopulated and repopulated by European colonial settlers in the sixteenth and seventeenth centuries. On most islands a Black African slave labour force was introduced, overseen by English, French and Dutch landlords, living in the 'great houses' of the day. The plantation system has been blamed for the long-standing underdevelopment of the Caribbean (Beckford 1972), and has persisted in a variety of forms and influences even after the decline of the agricultural produce it was intended to supply. Researchers have commented on the similarities in structure and form between the plantation system and mass tourism, including the dominance of non-local capital, control and markets (Hills and Lundgren 1977; Richards 1982).

Husbands (1983) draws this analogy further in a historic sense, in the context of the 'welcoming society', noting that 'the welcoming society was rooted in the social and political (and also economic) reality of the colony – it served a very deep-seated social and political function' (Husbands 1984:33). White visitors were a rarity in the early years of settlement and were entertained lavishly by hosts reaping the financial rewards of the boom in sugar. The subsequent sharp decline in sugar prices and drop in value of estates from the mid-nineteenth century onwards saw the gradual disappearance of much of the sugar production in many islands and the search for alternative economic activity.

One solution to the problem was to institutionalize the hospitality (Husbands 1984). Guests had previously been entertained and accommodated in the 'great houses' of the sugar barons, and travelled and lived among the infrastructure of the plantation system (Weaver 1988). It was relatively easy to convert some of these houses to hotels, and the land surrounding them to tourist-related developments. This pattern was not universally adopted in the region, as will be discussed below, but in several islands provided a relatively smooth transition from one export industry to another.

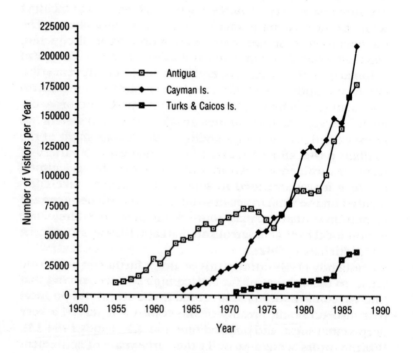

Figure 5.2 Visitor statistics

Source: See text.

The three cases selected for examination are Antigua, the Cayman Islands and Turks and Caicos. They are three of six island states included in a study of the process of tourism development in the Caribbean undertaken by the author with the assistance of graduate students in the period 1983–8. Detailed studies of Barbados (Husbands 1984) and Antigua (Weaver 1986) have been completed, and Weaver has also published specific papers on Antigua (1988) and the Cayman Islands.

The three states selected for discussion represent tourist destinations at different stages of development and at different scales. Antigua and the Cayman Islands are clearly further along the development process, as illustrated in Figure 5.2, than Turks and Caicos. The figures on visitor arrival (Table 5.2) alone, however, are a little misleading. Antigua had a recognizable tourist industry from the 1930s, with the first hotel being developed by 1908, and 205 steamers calling in 1929. By 1935 it was described as 'rediscovered' and by 1937 was a popular stop for cruise ships to the West Indies. The Cayman Islands, on the other hand, did not have a hotel until 1951, although their potential for tourism had been recognized in the early years of the twentieth century. The first commercial accommodation on Turks and Caicos did not appear until 1963, although ironically it now has the largest single development of any of the three islands, in the form of a Club Mediterranée complex.

In his paper 'The evolution of a "plantation" tourism landscape on the Caribbean island of Antigua', Weaver (1988) notes the relative ease with which tourism was accommodated in post-war years within the core-periphery structure of much of the Caribbean. Elements of similarity between tourism and the then declining plantation agricultural system include the importance of expatriate capital and entrepreneurial activity, including ownership and management of facilities, reliance on cheap unskilled local labour, a narrow market, seasonality and the focus of activity on external not internal needs. He notes that the end result has often been a tourist industry which perpetuates underdevelopment (ibid.:320). Weaver suggests that the decline in agricultural production, especially in sugar, was mirrored by a corresponding growth in tourism (ibid.:321). That relationship, however, is not necessarily a causal one

81

in a direct sense, for example, competition for land and resources. However, tourism has presented more attractive alternatives for capital investment and perhaps for employment of the local labour force. In Antigua plantation properties were converted into tourist developments as they became

Table 5.2 Stayover visitor statistics

	Antigua	Cayman Islands	Turks and Caicos
1955	10,038		
1956	11,602		
1957	13,309		
1958	16,134		
1959	21,009		
1960	30,221		
1961	26,512		
1962	35,101		
1963	43,272		
1964	46,118	4,834	
1965	48,651	6,626	
1966	55,657	8,244	
1967	59,174	10,278	
1968	55,838	14,460	
1969	61,262	19,411	
1970	65,369	22,288	
1971	67,637	24,354	2,742
1972	72,328	30,646	4,670
1973	72,786	45,751	5,881
1974	69,854	53,104	7,967
1975	62,971	54,145	8,138
1976	57,191	64,875	7,566
1977	68,297	67,197	7,693
1978	76,895	77,402	9,627
1979	87,759	100,587	9,739
1980	87,671	120,241	11,887
1981	85,824	124,598	12,348
1982	88,142	121,214	13,342
1983	101,113	130,763	14,216
1984	129,099	148,485	17,263
1985	140,000	145,000	30,185
1986	166,200	166,082	35,000
1987	186,700	209,044	36,600
1988*	213,500	164,500	32,000
1989*	220,000	181,000	36,000

Source: Caribbean Tourism Research Centre (1991) *Visitor Statistics*,
Barbados: CTRC.
Note: *Preliminary figures.

financially unattractive in agriculture, and some 'great houses' remodelled into hotels. Thus the plantation system, its capital and its owners provided a nucleus of resources, physical, financial and human, from which tourism development proceeded at a rapid pace in the 1960s and 1970s in Antigua (Weaver 1988), following the inauguration of jet aircraft service with the USA in 1959. One element is perhaps at odds with the cycle in the case of Antigua. In the model the early stages are characterized by local ownership and control of development. It could be argued that the expatriate involvement in tourism in Antigua in its early stages is not local in the sense of indigenous people, nor is the orientation at all local in focus. This is a moot point, as some expatriate families predate or are contemporary with slave settlement and certainly view themselves as Antiguan. The progress of Antigua through the cycle has not been smooth. The rapid rise in oil prices in the early 1970s and the mini depression of the early 1980s saw what appears to be premature declines in growth, as visitor numbers rebounded in the late 1970s and again in the mid-1980s. Some problems relating to the environment have already appeared in Antigua, along with seasonal overuse of resources, and the non-local control of resources is illustrated by the fact that foreign business interests are estimated to control 90 per cent of Antigua's accommodation (Weaver 1988:328).

The Cayman Islands represent a destination in a somewhat earlier phase of development than Antigua, experiencing more rapid rates of growth in numbers of visitors, and still in the middle development stage of the cycle. There is little specific information on tourism in the Caymans prior to the 1960s. As noted above, no hotels existed before 1950, and in the years prior to this visitors had to stay as guests of locals or in government quarters. A simple local inn and beach cottages were gradually changed through the 1950s and 1960s, and a key to the appearance of tourists in increasing numbers was the improvement of transportation services. As Weaver notes (1990), the construction of the airport on Grand Cayman was due to a London investment group who also planned to expand tourism, and resulted in a rise in real estate values, representing a measure of non-local intervention from a remarkably early stage in the cycle. In a decade land values increased some

400-fold in specific locations, and hotel accommodation rose from zero to over 300 beds in the decade from 1951 to 1961.

In the decade 1971 to 1981 extensive development occurred, matched by a more than 500 per cent increase in staying visitors, while cruise ship tourists increased from 1,000 in 1972 to 158,000 in 1982. A heavy emphasis on the US market (> 70 per cent) emerged, and extensive marketing has taken place. Accommodation has expanded in a similar way and condominium and self-catering is now increasing more rapidly than hotel development. Physical changes have occurred, including golf course development and redevelopment in George Town, the capital. Air links to major markets have been improved and cruise ship facilities have been constructed.

While in many respects the nature and pattern of development closely resembles the model proposed, Weaver (1990) notes some significant differences. The major one relates to the almost total absence of local involvement in the early stages of involvement and development in the Caymans. External interests were responsible for the first resort hotel, the airport and land development and speculation (ibid.). The lack of local capital and entrepreneurs may be due in part to the lack of a plantation agriculture system in the Caymans and also to dependency status. No 'welcoming society' as existed in plantation islands such as Antigua and Barbados was present in the Cayman Islands to be institutionalized or provide a catalyst to tourism development. As Weaver notes, however, local involvement in tourism appears to have begun later in the Caymans and actually increased during the development stage rather than declining as suggested in Table 5.1. This is reflected in some local investment, and especially in the imposition of locally-established planning and regulation of development. The latter may be a result of increased political autonomy and the former because of a relatively high local level of income based not only on tourism but also on the Cayman's success as an offshore financial centre. Although the Cayman Islands are still experiencing rapid development and thus still have a considerable way to progress along their cycle, and display some significant differences from the model, Weaver concludes that both [Antigua and Cayman Islands] *do* [his emphasis] conform in many critical respects with the model (ibid.).

In contrast to the examples above, tourism in Turks and Caicos is much less developed and at an earlier stage in the cycle. Statistics on visitor arrivals are only available from 1971, when less than 3,000 staying tourists came to the islands. The first formal hotel was constructed in 1963, and a tourist board established in 1972. The islands are perhaps the least developed of any of the British dependencies in the Caribbean, and development potential is limited. A small plantation economy had existed but ceased early in the twentieth century and the major source of revenue was the export of salt, which existed until the 1960s. The economy in the post-war years has relied heavily upon subsidies from the British government and tourism has been seen to be the only identifiable key to economic development, especially with the closure of most of the American defence and space exploration facilities in the last decade (Butler 1988).

In a pattern similar to that of the Cayman Islands, the impetus for tourism development came from expatriates and visitors to the islands, rather than from local initiatives. The main tourist developments, although not the earliest, which took place on South Caicos, have been on the island of Providenciales, and were initiated by Americans. In exchange for road and airstrip construction, a small American consortium was granted a large tract of Crown land, which proved the location for the first two hotels in the mid-1960s. The proximity of the islands to the United States and to the Bahamas has been a factor in shaping the visitation pattern. Yachtsmen based in Florida and the Bahamas became familiar with the islands, despite the rather dangerous conditions in many locations, and air links to Florida, if irregular and spasmodic, were attractively short in duration. The American bases in Grand Turk provided further links with the United States. A high proportion of visitors, however, have always been Canadian, and in the 1970s and 1980s various unsuccessful efforts were made to formally link the islands with Canada, including a petition by the islands to the British government to be transferred to Canadian jurisdiction in 1974. This probably reflects the annual onset of 'cabin fever' among many Canadians, and desperation for economic development among islanders, rather than economic and political reality.

Visitor numbers to Turks and Caicos reflected the gradual rise in accommodation capacity, visibility and accessibility from the 1960s onwards, but while development levels corresponded to the exploration and involvement stages of the cycle until the 1980s, almost all of the development was owned and operated by expatriates, recent immigrants or absentee landlords (Butler 1988). Although in relative terms numbers of tourists rose sevenfold from 1971 to 1984, the total figure was still well below 20,000 visitors a year (Table 5.2). The first significant step into the development stage came in 1984 with the completion of a Club Mediterraneé development of 300 rooms in Providenciales. This development took several years and significant concessions to secure, including the development of an international airport paid for by the government, a surfaced road to the Club site, and taxation and water-supply arrangements. The effect of this development on visitor numbers was dramatic, with an increase of almost double within one year. The presence of the Club and the airport also resulted in improved air services from the United States, although not consistent over the period 1985 to present, and from Canada in 1987. Coopers and Lybrand (1986) note the feeling on the island that one major development was needed to trigger development, although they rightly question whether this would guarantee development in the absence of a complete infrastructure and the number of islands in very different stages of development. While Providenciales is now in the development stage, the remaining islands are at best in involvement (Grand Turk) or entering that stage (North and South Caicos), a source of some inter-island rivalry and resentment.

There is no doubt that development may proceed rapidly, at least on Providenciales, in the decade ahead. A review of planning applications and decisions relating to development over the period 1975–85 shows accelerated demand (43 applications from 1975-9, and 189 from 1980 to 1985), but as might be expected, despite a generally high level of approval, there was a low (about 28 per cent) rate of completion of projects. The rate of completion is lower for the second half of the period (26 per cent) than for the first half (38 per cent) (Butler 1988). Almost all of the proposed development has been supported by overseas capital and can be viewed as international. Local involvement, except at a limited scale, and then monthly by recent immigrants, is confined to minor services and not related to major developments. This reflects primarily the absence of an entrepreneurial class in the islands with sufficient capital to initiate major developments, perhaps again reflecting the absence of a strong plantation-based economy with characteristics noted above in the context of Antigua. While Connell (1988) notes that small islands may benefit from strong links with a major economic power as, for example, in a dependency situation, this does not seem to have aided Turks and Caicos in breaking into the early stages of development. However, the presence of a planning and development department, and the use of consultants, may have served the islands well in the long term, by curtailing, or at least delaying, some potentially unwise and inappropriate, at least in terms of scale, developments, and have certainly aided the protection of the natural resources and environment. It will be some years before it will be possible to judge the true effect of the institutional arrangements on tourism developments in the islands. Further development in Turks and Caicos seems certain, but some individual islands, because of relative inaccessibility to markets, small size, and small population, may remain little developed and see only small resort or self-catering accommodation developed, which may well be much the most appropriate pattern. The development pattern of the Hawaiian Islands may prove illustrative as a comparison and an indication of the future of Turks and Caicos, with Oahu representing Providenciales, albeit at a much larger scale, for example, Maui for Grand Turk, and Hawaii for North Caicos. Thus while this destination has followed the pattern illustrated in the model as regards numbers of visitors, the influences on development, as in the Caymans, indicate some variation from those hypothesized.

FUTURE TRENDS

At the regional scale there is little doubt that tourism will continue to grow in the Caribbean basin. In the first instance, the region has a tremendous locational advantage relative to the North American market which has always provided the majority of its visitors. This locational advantage may be strengthened economically if fuel costs rise, making longer-haul vacations from North America proportionally more expensive. The climatic attraction, coupled with landscape and cultural appeal, and with English as the primary language make this region highly attractive to North Americans. Expatriate tourism can also be expected to rise as increasing numbers of Caribbean emigrants, especially, but not only, those in North America, achieve a level of affluence sufficient to allow them to visit their homelands.

Specific events may alter this pattern slightly. If Cuba renounces Communism, or enters into a less-negative relationship with the United States, visitation to this island may attain new levels (Hinch 1990). Political events in Eastern Europe, however, may result in a decline in Cuban tourism from that area, as other destinations closer to Eastern Europe are now accessible to Eastern Europeans. Independence for the remaining dependencies in the Caribbean could see additional rapid development if planning and development controls are eased or changed. Political instability in any island on the other hand, could easily result in a decline in tourist numbers, as Jamaica and Grenada have experienced in the past, and possibly Trinidad and Tobago in 1990–1. Haiti, in recent years and months, has suffered more than any other destination in the region because of its political situation. Such events can clearly be of a magnitude sufficient to alter a destination's progression through its cycle, delaying development or hastening or triggering decline as the case may be.

One may make some general forecasts for the three cases briefly discussed. In the short term, one destination, Turks and Caicos, is likely to proceed fully into the development phase, a second, the Cayman Islands, will continue in this phase, and the third, Antigua, may enter the stagnation phase. In the medium term, Turks and Caicos should continue its development, but the Cayman Islands are likely to enter consolidation by the end of this decade, as much from running out of physical space as any other factor. Antigua, in the medium term, is likely to see increasing conversion from hotels to condominium/time-share type developments and fewer conventional hotel developments. Numbers of tourists should stabilize in the medium term, with the exception for all areas of continued increases in expatriate tourism. Under present circumstances much of this tourism is likely to be in the current 'off-season', that is, the summer, when prices and demand are lower than in the winter. This pattern too may change over time. As in some European destinations, increasing attention is likely to be paid to attracting long-stay shoulder-season visitors, particularly the growing elderly segment of the market. The limited nature of accommodation, accessibility and attractions will remain a constraint on Turks and Caicos until a critical mass of development occurs which will allow regular and frequent air service and the tapping of the mass market. Such may not be desired by the authorities there, but is certainly the desire of many of the current developers (Butler 1988). The small population of the islands will limit expatriate tourism and the absence of attractive harbours limit cruise ship and yacht traffic, at least in the short term.

CONCLUSIONS

The emphasis in the paper has been on a descriptive explanation of the pattern of development of tourism in islands, particularly in the Caribbean. The general model of development proposed provided a background against which factors influencing development were discussed. It is felt that the model withstands examination in these instances reasonably well, with some expected variation. This conclusion is shared by a number of other researchers (Edwards 1988; Weaver 1988; 1990; Wilkinson 1987). It has been suggested that at least some of the variation which does occur between the island states, and between the model and the examples, stems from the unique historical development of the islands, especially the presence or absence of plantation agriculture. Other factors affecting tourism development have included political status, language, population, general economic level of development and location within the Caribbean. In the future, the legacy of the past will be of less significance in ensuring competitiveness and survival in the tourist industry than will control and management of the tourist product and its quality, and realistic and accurate marketing of truly unique features of the destination. Islands which do not succeed in this regard will decline in tourist appeal and will be indistinguishable from each other or, in Relph's (1972) terminology, placeless.

REFERENCES

Beckford, G.L. (1972) *Persistant Poverty: Underdevelopment in Plantation Economies of the Third World*, New York: Oxford University Press.

Britton, S. and Clarke, W.C. (1987) *Ambiguous Alternatives: Tourism in Small Developing Countries*, Suva: University of the South Pacific.

Bryden, J.M. (1973) *Tourism and Development*, Cambridge: Cambridge Univesity Press.

Butler, R.W. (1980) 'The concept of a tourist area cycle of evolution: implications for management of resources', *Canadian Geographer* 24(2): 5–12.

Caribbean Tourism Research Centre (1991) *Visitor Statistics*, Barbados: CTRC.

Cohen, E. (1972) 'Towards a sociology of international tourism', *Social Research* 39(1): 164–82.

Connell, J. (1988) 'Contemporary issues in island micro-states', in M. Pacione (ed.) *The Geography of the Third World*, London: Routledge, 427–62.

Coopers and Lybrand Assoc. (1986) *Turks and Caicos Islands Tourism Development Plan*, London: Overseas Development Administration.

Edwards, F. (ed.) (1988) *Environmentally Sound Tourism in the Caribbean*, Calgary: University of Calgary Press.

Hills, T.L. and Lundgren, J. (1977) 'The impact of tourism in the Caribbean: a methodological study', *Annals of Tourism Research* 4(5): 248–67.

Hinch, T. (1990) 'The Cuban tourist industry: its re-emergence and future', *Tourism Management* 11(3): 214–226.

Holder, J.S. (1979) *Caribbean Tourism Policies and Impacts*, Barbados: CTRC.

Husbands, W.C. (1983) 'The genesis of tourism in Barbados: further notes on the welcoming society', *Caribbean Geography* 2: 107–20.

Husbands, W.C. (1984) *Pattern, Structure and Formation of Activity Space in Hinterland Resorts: A Study of Barbados*, PhD thesis, London: Department of Geography, University of Western Ontario.

Meyer-Arendt, K.J. (1985) 'The grand-isle Louisiana resort cycle', *Annals of Tourism Research* 72(4): 449–65.

Pacione, M. (1988) *The Geography of the Third World*, London: Routledge.

Pearce, D. (1980) *Tourism in the South Pacific*, Christchurch, NZ: Department of Geography, University of Canterbury.

Pearce, D. (1987) *Tourism Today: A Geographical Analysis*, New York: Longman.

Plog, S.C. (1972) 'Why destination areas rise and fall in popularity', paper presented to Travel Research Association, Los Angeles.

Rajotte, F. (ed.) (1982) *The Impact of Tourism Development in the Pacific*, Peterborough: Trent University.

Relph, E. (1976) *Place and Placelessness*, Toronto: Abacus.

Richards, V.A. (1982) 'The Antigua and Barbudan Economy: Trends and Prospects', *Antigua and Barbuda Forum* 1: 30–6.

Sadler, B. (1988) 'Sustaining tomorrow and endless summer: on linking tourism and environment in the Caribbean', in F. Edwards (ed.) *Environmentally Sound Tourism in the Caribbean*, Calgary: University of Calgary Press, ix–xxiii.

Seward, S.B. and Spinard, B.K. (1982) *Tourism in the Caribbean: The Economic Impact*, Ottawa: IDRC. (International Development and Research Centre).

Stansfield, C. (1978) 'Atlantic city and the resort cycle', *Annals of Tourism Research* 5(3): 238–51.

Weaver, D.B. (1986) *The Evolution of a Heliotropic Landscape: The Case of Antigua*, PhD Thesis, London: Department of Geography, University of Western Ontario.,

Weaver, D.B. (1988) 'The evolution of a "plantation" tourism landscape on the Caribbean island of Antigua', *Tijdschrift voor Economische en Sociale Geografie* 29(5): 319–31.

Wilkinson, P.F. (1987) 'Tourism in small island nations: a fragile dependence', *Leisure Studies* 6(2): 127–46.

Wilkinson, P.F. (1988) *Tourism in Island Microstates: Bibliography*, Monticello, Illinois: Public Administration Series: Bibliography P2504, Vance Bibliographies.

6

BASIC NEEDS AND DEVELOPMENT IN THE SMALL ISLAND STATES OF THE EASTERN CARIBBEAN

Robert B. Potter

INTRODUCTION

This account assesses the extent to which the small island states of the Eastern Caribbean have been able to establish more self-reliant indigenous strategies of development in the post-independence era. At the outset, it is fully acknowledged that any such evaluation must take into account the multiplicity of constraints to development that are faced by these small island microstates. In particular, the small territorial size, fragile ecological and narrow resource bases, diminutive populations, and the highly-dependent nature of the economic structures of these territories are factors which have to be kept to the fore in any discussion of their development paths.

The chapter focuses in particular on the small impoverished states of the Windward Islands, specifically Dominica, St Lucia, St Vincent and the Grenadines, and Grenada. In appropriate places reference is also made for comparative purposes to the more developed Eastern Caribbean territories of Barbados, and Trinidad and Tobago. The main body of the text considers the degree to which development strategies, as they have involved the agricultural, manufacturing, retail, tourism and housing sectors, have endeavoured to serve the basic needs of the population in these countries. The materials on basic needs and the provision of shelter draw specifically on the author's current research into housing conditions and state policies in the Windward Islands. The basically centre-right development ideologies adopted by the majority of these states is clearly highlighted in this overview of economic and social change in the post-independence era. Linked to this, a simplified model of small island Caribbean development is presented. Subsequently, the topical issues surrounding the current proposal for the political union of the Eastern Caribbean states of Dominica, St Lucia, St Vincent and the Grenadines, and Grenada are considered.

In the concluding section, specific attention is accorded to the development policies pursued by the People's Revolutionary Government (PRG) in Grenada between 1979 and 1983. This was the first avowedly socialist government within the Commonwealth Caribbean, and one which made direct and explicit efforts to promote self-reliance in agriculture and to provide for the basic needs of the populace at large.

THE BASIC CHARACTERISTICS OF THE EASTERN CARIBBEAN SMALL ISLAND STATES

The four islands making up the Windwards, which form the prime focus of the present account, are shown in Figure 6.1. This group of island forms a broad arc running from Dominica in the north to Grenada in the south. Barbados lies approximately one hundred miles to the east of St Vincent, and to the north of the Windwards lie the Leeward Islands, consisting of Anguilla, St Kitts-Nevis, Montserrat, and Antigua-Barbuda.

These islands are extremely poor by world standards, indeed they are normally cited as the poorest territories in the Caribbean, with the exception of Haiti. In 1987 their Gross Domestic Products stood at Dominica US $1,440, Grenada US $1,340, St Lucia US $1,370, St Vincent and the Grenadines US $1,070 (Table 6.1). Each of the islands is small, ranging between the 790 km² of Dominica, and the 344 km² of Grenada (Table 6.1). The islands are ruggedly mountainous, being volcanic in origin.

Although the tourist and construction sectors of the economy have grown quite rapidly over the past few years – a trend which will be considered fully later in the chapter – agriculture, in particular, the production of bananas, remains

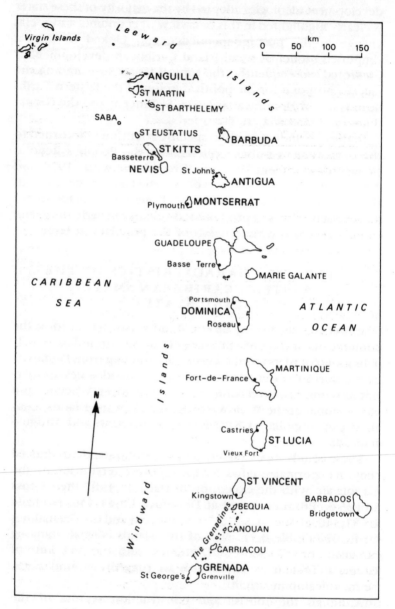

Figure 6.1 The Eastern Caribbean

Table 6.1 Basic data for the Windward Islands

Country	Population (mid-1987)	Gross National Product pc 1987 (US $)	Population growth (% pa 1980-7)	Life expectancy (1987)	Area (km²)
Dominica	80,000	1,440	1.3	75	790
Grenada	100,000	1,340	2.0	69	344
St Lucia	143,000	1,370	2.1	72	617
St Vincent	112,000	1,070	1.3	69	389

Sources: The World Bank Atlas, 1988; Population Census of the Commonwealth Caribbean (1980-1).

the mainstay of these economies. Indeed, agricultural cash crop exports generally provide between 60 and 85 per cent of the area's total exports and account for between 20-40 per cent of total GDP (Fraser 1985). On the other hand, industry is in its infancy, currently accounting for less than 10 per cent of total GDP among the islands, although again, this is a sector of the economy in which quite rapid progress is starting to be made. Despite the high rainfall totals which are attendant upon the marked relief of the islands, and the preponderance of black volcanic derived beaches, tourism grew as an industry during the 1970s and 1980s, particularly in St Lucia. Thus, it is now estimated that tourism employs almost one quarter of the labour force of the Windward Islands, and that it provides between 50 and 70 per cent of the area's foreign exchange earnings (Fraser 1985).

The islands of the Eastern Caribbean are all heavily dependent on imports, and in 1987 three of them exhibited substantial balance of payments deficits (Table 6.2).These ranged from Bd $25.6 millions in the case of St Lucia, to Bd $35.5 millions for Grenada (BD $2 = US $1). Only Dominica recorded a surplus, at a mere Bd $8.4 millions. It can be argued that the expansion of tourism has done little to help in this regard, leading to substantial increases in imports, both of capital goods for the development of infrastructure and for hotel construction, and also of foodstuffs (Fraser 1985). Saliently, as Table 6.2 readily demonstrates, for all four territories, the total value of imports is approximately two times greater than the value of exports. As a direct consequence, countries like Grenada have increased sharply their external borrowing to finance the balance of payments deficit, so that external debt rose from US $58.4 millions in 1986 to US $69.4 millions in 1987, the latter figure being equivalent to 46 per cent of GDP. Similarly, in St Vincent and the Grenadines the external public debt amounted to some 25 per cent of GDP in 1986.

Table 6.2 Balance of payments statistics for the Windward Islands 1985 and 1987 (provisional) (Bd $millions)

| | *Dominica* | | *Grenada* | | *St Lucia* | | *St Vincent* | |
	1985	*1987*	*1985*	*1987*	*1985*	*1987*	*1985*	*1987*
Imports	110.7	111.6	138.5	176.7	250.0	356.2	158.4	197.3
Exports	56.9	86.8	44.7	70.6	104.0	154.6	126.5	104.7
Service balance	4.4	3.0	21.4	28.4	81.0	126.6	15.4	36.6
Net transfers	42.6	30.2	80.7	42.2	46.2	49.4	34.2	28.5
Balance on current account	−6.8	8.4	8.3	−35.5	−18.8	−25.6	17.7	−27.5

Source: Central Bank of Barbados (1989).

Even this brief overview of the geography and economy of these small island countries is sufficient to point to the constraints which they face, as well as giving some indication as to the sorts of development plans which have been adopted by them in the past. Before considering in detail the paths to development followed in these nations since independence, the nature and definition of basic needs will be considered, specifically as they relate to the circumstances that are to be found in the Eastern Caribbean region.

BASIC NEEDS AND THE EASTERN CARIBBEAN REGION

By the middle to late 1970s there was increasing agreement that the basically economic-growth-oriented strategies pursued in Third World countries as a whole during the 1950s and 1960s had done little to alleviate poverty (see Hettne 1990). Even where moderate growth rates had been recorded nationally, the benefits had been far from equitably distributed (Potter 1991a). The proponents of trickle-down growth themselves had to admit that the process was at best exceedingly slow. As a logical reaction, a growing school of developmentalists proposed that a direct approach was needed whereby the express aim of development was to reduce poverty. The approach effectively argued for redistribution rather than economic growth on its own, and was referred to as the basic needs strategy.

The idea of basic needs is in many ways not new, and further, as recently noted by Hettne (1990), in several respects is self-evident. Thus, many countries that followed people-oriented approaches to development planning have subscribed to elements of the basic needs approach without ever using the term itself. The need for an approach which emphasizes needs satisfaction was strongly advanced by the International Labour Organization (ILO) from 1969 (see ILO 1976a and 1976b). Specifically, their recommendations involved the promotion of employment growth in the agricultural and informal sectors, with this taking priority over the pursuit of economic growth.

Leaving aside the complex debates which surround the nature and precise definition of what constitute the basic needs of a given population or cultural group (see Hettne 1990: 168–70 on this), the approach clearly involves considerations of the minimum standards of living which a society should set for the poorest groups of its population (ILO 1976a). Frequently, the need to 'feed, house and clothe the nation' is cited as its prime aim. But more widely cast, the net of basic needs encompasses the provision of services such as safe water and sanitation, transport, health care and education, as well as the maintenance of basic human rights such as liberty, equality, justice and effective participation in the decision-making process.

It is clear from the foregoing that basic needs provision should be seen as an overarching national development strategy, and not as a series of palliative measures. As such, the approach is connected with the quest for greater self-sufficiency. More particularly, the argument that the needs of the poor should be met in priority to externally-oriented growth imperatives is central to basic needs provision. Thus, the approach can be broadly aligned with the rejection of excessive exogenous influences as a part of the process of convergence on western patterns of consumption in respect of diets and consumer behaviour (Macleod and McGee 1990; Drakakis-Smith 1991 and Potter and Dann 1991), as well as in other spheres. Other things being equal, exogenous influences basically serve the non-basic needs of the middle and upper sectors of society.

Couched in these general terms, the implications of this type of approach are potentially very interesting for territories such as those found in the Eastern Caribbean. First, these countries became independent at the end of the 1970s, at the very time when basic needs approaches were being championed internationally. Given that the whole history of the Caribbean has been intimately associated with European colonialism, and bearing in mind the subsequent economic, socio-cultural and psychological dependency that this has engendered (Beckford 1972), the adoption of more self-reliant, autonomous and indigenous paths to change and development stands as an important corrective in the region. In the past, for example, such territories have been involved in producing agricultural staples for export, whilst as a direct corollory, vast quantities of day-to-day foods and maufactures have had to be imported. Thus, Downes (1980) in one of the few papers written on the provision of basic needs in the Eastern Caribbean, argues that it should afford these microstates the opportunity to match their development paths to their own particular environments and needs. The approach is also of salience when it is recognized that the extremely small size of the islands means that the comprehensive adoption of agropolitan principles of development, in association with selective regional closure, is not realistic for these territorires (Potter 1989a). It can be argued that even Cuba has merely replaced capitalist economic dependence, for socialist economic dependence (Hall 1989). Although closely associated with a number of principles of agropolitan development, it is argued here that in the case of small countries basic needs strategies are more easily integrated with enforced exogenously-oriented paths to development, via the imposition of mechanisms of social and spatial redistribution.

PATTERNS OF ECONOMIC CHANGE IN
THE EASTERN CARIBBEAN

Just after the period of granting of independence, Downes (1980) noted that the less-developed countries of the Windward and Leeward Islands had generally followed in a passive manner the progression of approaches to development pursued previously by their more developed neighbours such as Jamaica, Trinidad, Barbados and Guyana. Given their peripheral dependent capitalist status, the conventional wisdom followed by these small island Eastern Caribbean nations has been that of maximizing economic growth via export-led agriculture and industrialization by import-substitution and then, by invitation of enclave enterprises.

Thus, relatively little attention has been given to agriculture for domestic purposes. For centuries, agriculture has dominated the economies of these islands, based at least in part on the plantation system and with high levels of foreign ownership. The plantation system has been associated with a pattern of persistent poverty by the Jamaican social scientist George Beckford (1972). In the Eastern Caribbean islands, since emancipation, there has been a quite substantial smallholder or peasant agricultural sector (Shephard 1947, Rojas, 1984). Indeed, the decline of the former plantation-produced staple sugar, and the introduction of the replacement staple of bananas, has given rise to a number of land redistribution schemes on former plantations in both the Windward and Leeward Islands (Rojas and Meganck 1987). However, both the large- and small-scale agricultural sectors have continued to contribute mainly to export staple production since independence. Agricultural productivity has decreased for the region taken as a whole and many countries are now heavily dependent on food imports (Gumbs 1981). Long-established systems of land tenure such as those based on the French system of equal inheritance in St Lucia, and continuing export crop monoculture, are two of the factors frequently seen as being involved. Others include ongoing processes of soil erosion and land degradation.

As previously noted, today in the Eastern Caribbean the main export crop is bananas, and what is seen as the contribution of this crop to the continuing poverty and dependency of the region has recently been explored by Thomson (1987), who refers to bananas as 'green gold'. In a good year, the crop can be a licence to print money for the British Company Geest, who are the sole importers and marketers of Windward Island bananas into Britain. Yet for the individual small farmer, production is an extremely risky business, not least because of the potential damage that can be done by hurricanes, such as Allen in 1980. Thus it is argued that both poverty and dependence continue long after political independence has been achieved. The essential corollary is that, despite some limited attempts by government to increase agricultural diversification in the Eastern Caribbean, the basic nutritional needs of the population are frequently only being met by the costly importation of foodstuffs. This is exemplified for the Eastern Caribbean territories by the data contained in Table 6.3, which show that food imports amount to around 30 per cent of all imports. Items such as tinned meats and fish, rice, biscuits and flour are typically imported. There is also a heavy reliance on manufactured imports, as shown by Table 6.3. Clearly there remains a pressing need for the introduction of more efficient modes of using agricultural land, along lines which will encourage self-sufficiency and sustainability.

Table 6.3 Food and manufactured goods as a proportion of total imports into the Windward Islands

	Percentage of total imports		
	Food and live animals	*Beverages and tobacco*	*Manufactured goods*
Dominica	22.0	3.2	29.2
Grenada	28.2	2.8	23.0
St Lucia	20.4	2.8	34.2
St Vincent	27.5	4.4	28.5

Source: Fraser 1985, from Organization of Eastern Caribbean States.

As Connell (1987) has commented, post-independence development strategies in small island states have generally been evolutionary and not revolutionary. Thus, despite espousing the need for greater self-sufficiency and the better use of domestic resources, this has rarely happened in practice.

This has been as true in other sectors of the economy as it has in connection with agriculture. The promotion of light manufacturing along the lines advocated by Sir Arthur Lewis (1955) has occurred, generally involving industrialization by invitation. A local model for this was provided by Barbados and its 'Operation Beehive' launched in 1969 (Potter 1980). This provided for tax holidays and the duty-free importation of all materials for overseas companies establishing branches on one of the nine industrial estates constructed by the Industrial Development Corporation. By 1980 manufacturing accounted for 12 per cent of Barbadian GDP, and provided 9,000 jobs. Although in their infancy, both St Lucia and St Vincent and the Grenadines have followed similar paths, focusing respectively on their Hewanorra and Camden Park industrial estates. Harrison (1984) has noted that, by the early 1980s, manufacturing contributed 12 per cent to the GDP of St Vincent, this having doubled in a mere six years.

However, there are obviously likely to be costs as well as economic gains involved in programmes which are based on foreign ownership and control. Operation Beehive, for example, involved the establishment of some 173 transnational corporation branch plants in Barbados, and Kowalewski (1982) argues that such companies promote the interests of elite groups, whether these are local or overseas, thereby diverting attention away from the pressing needs of the poor. A major problem is that when the tax concessions run out, or profitability is otherwise reduced, multinationals are likely to move swiftly and without ceremony in closing branch plants, without having to give any thought to the social consequences of their actions. Additionally, it is argued that such developments, by their very nature, are based on the creation of lowly-paid jobs, mainly for women, and they may well involve very poor working conditions, as noted by Kelly (1986) for St Lucia (see also Barry, Wood and Preusch 1984). Further, in connection with the policy of industrialization by invitation in St Vincent, Nanton (1983) notes that two firms actually persuaded the government to break its own minimum-wage legislation by demanding and obtaining a 'training rate' at 30 per cent less than the statutory minimum wage then in force. Kelly (1986) also points to the fact that in St Lucia many of those working in factories do so in order to acquire skills which they hope will subsequently help them to emigrate to metropolitan countries. Many of these arguments apply equally well to the new growth area – that of offshore data processing. In all these ways, the paths followed in the post-independence era have done little to serve the basic needs of the poor in these territories. Suggestions that emphasis should be placed on the creation of rural and informal sector jobs are likely to be met locally either with indifference or open disapproval. Certainly, it is hard not to reach the conclusion that the informal sector has received surprisingly little academic and policy-oriented attention in the Caribbean (see Lloyd-Evans and Potter 1991). In these ways, external rather than domestic needs have once again been served before those of the local populace.

Much the same types of argument can be advanced in respect of that other growth sector of the economy – tourism. St Lucia is the major attraction for stop-over visitors (78,700 in 1983), whilst Grenada has traditionally been the favourite port-of-call for cruiseship passengers (50,200 in 1983). Governments in the various territories have offered fiscal and financial incentives to encourage hotel construction (see Potter 1983). With regard to the potential development impact of tourism, it has been estimated that as much as 70 per cent of all tourist expenditure in the Caribbean region is repatriated, principally due to the very high level of foreign ownership of hotels (Fraser 1985). Thus, in 1970, eleven of St Lucia's fifteen hotels were owned by North Atlantic transnational corporations. Nanton (1983) records that whilst by 1974 tourism provided 400 jobs in St Vincent, 60 per cent of the hotels on Bequia were US-owned. For the same year, Nanton (1983) notes that in St Vincent 65 per cent of employment and over 70 per cent of hotel income were generated by ten foreign-owned hotels. Similar statistics are recorded throughout the Caribbean, so that in Barbados in 1980, some 74 per cent of Class 1 hotels were owned by foreigners (Potter 1983).

In these circumstances, it becomes clear that tourism as an explicit development strategy is unlikely to contribute to basic needs. This is also the case because most evidence points to the fact that its growth often exacerbates dependence on imports, especially those of foods. This occurs not only by means of the demands of the tourist palate itself, but also due to the tourist demonstration effect. Equally, much evidence suggests that the flow of workers to the tourist sector dampens the availability of labour for agriculture, and may lead directly to the creation of idle land. The competitive bidding up of land prices in the tourist-oriented small islands of the Eastern Caribbean is also well-documented (Potter 1989b). Restrictions may well be placed on locals if large tracts of land, or indeed entire islands, are purchased by foreigners. In the Grenadines, for instance, both Mustique and Petit St Vincent were purchased by overseas companies, whilst Mayreau is owned by a local landlord. Further, 17.5 per cent of Bequia is owned by one US company. In the case of Mustique, the local population were not allowed to own property and their birth and burial rights were curtailed in case they should claim rights to the land (Nanton 1983).

The account would not be complete without reference to the other great source of income, remittances from nationals living overseas, a substantial occurrence in the Eastern Caribbean. If development is defined as the move away from dependency on the external environment, toward greater autonomy and self-reliance, the past and continuing emigration of the young and able must be regarded as a powerful constraint on progress (Marshall 1982). Further, most authors who have looked at the effects of remittances have argued that they generally have a deleterious effect on agricultural production. Momsen (1986) notes that in St Vincent, frequently, money earned overseas is used for investment in land for retirement purposes, which consequently often remains idle for lengthy periods. Also in relation to St Vincent, Rubenstein (1983) has argued that reliance on remittances both to finance schooling and for the purchase of highly-valued imports only further intensifies the desires of others to migrate. All of these developments are clearly antipathetic to basic needs-oriented development for the poorest sectors of society.

In several recent papers (Potter, 1989a, 1989b), the present author has tried diagrammatically to summarise some of the consequences of these various strands of externally-oriented, growth-maximising, highly dependent paths to development and change in the small island states of the Caribbean. The resulting diagram is shown in Figure 6.2 and is presented here as a generalised model of the structure of very small island economies. The first point is that such small islands seem to have become more rather than less economically dependent since their political independence. To the old established forms of dependency must be added those of imported foods, enclave industries, consumerism in general and new technology in particular, such as that offered by microcomputers. The spatial pull of existing urban zones and coastal areas has served to concentrate yet further new developments into these restricted, highly-developed localities. The channelling of new developments into such relative economic oases has left economic deserts. In such small territories the contrasts between the two areas is maintained by the movement of the rural poor to the urban and coastal zones for all manner of even quite basic activities and services, as is detailed in the diagram. Indeed, in very small island states, small size may be used as a reason justifying continued concentration in the future as a matter of public policy. Whilst this may or may not be economically optimal (Rojas 1989), the social costs are bound to fall on the rural and the poor, and are likely to be considerable, as Figure 6.2 implies.

As Connell (1987) records, the general response to the development problems of small island states is to pursue some form of economic integration with their neighbours. In the post-independence era such states have effectively become integrated with the economies of the metropolitan powers. Despite the development of CARIFTA (Caribbean Free Trade Area) in 1968 and its replacement with CARICOM (Caribbean Common Market) in 1973, and the existence of OECS (Organization of Eastern Caribbean States), these small nations still effectively stand alone. As long ago as 1968 O'Loughlin, in her study of the social and economic development of the Windward and Leeward Islands, commented that underlying the need for development is the search for a new form of nationhood, a theme later reiterated by Gordon-Somers (1984). In the past two years a serious proposal for the creation of a single state made up of Dominica, St Vincent, St Lucia and Grenada has been considered in the region. The driving force behind the proposal is the common concern over the fragile nature of the small economies of the separate nations. These states are proposing to carry out referendums on the issue early in the 1990s. If the creation of a single state

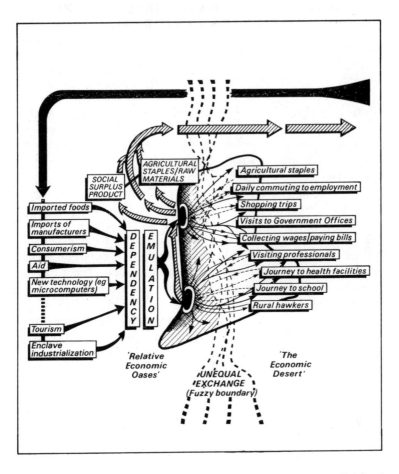

Figure 6.2 A generalized model of the structure of very small island economies in the Eastern Caribbean

would afford the strength and confidence to balance more finely the emphasis that is placed on external and internal agencies of change and development, then such political union would undoubtedly be a step in the right direction. As the foregoing account has sought to demonstrate, in the post-independence period few if any real efforts have been made to provide for the basic needs of the poor.

HOUSING AND BASIC NEEDS IN THE EASTERN CARIBBEAN

Another area where basic needs are of direct relevance is in respect of housing. The author is currently working on a project investigating housing conditions and state policies in the Windward Islands of St Lucia, St Vincent and Grenada (Potter 1991a, 1991b, 1991c), following a similar project in Barbados (Potter 1989c, 1992). People in the region have long since taken the responsibility for the construction of their own houses, traditionally small timber cabins. In 1980, in St Lucia, 73.62 per cent of houses were still constructed entirely of wood. The equivalent figures were 60.74 per cent for Grenada and 45.29 for St Vincent. In addition, for each island, approximately 40–50 per cent of houses are over twenty years old. These figures represent a considerable burden for maintenance by the poor. For each of the three territories, between 50 and 62 per cent of all dwellings use pit latrine toilets. The incidence of public standpipes for water supply remains surprisingly high, between 34.21 per cent in the case of Grenada and 45.33 per cent for St Vincent (Figure 6.3). This variable, in common with the others, shows a marked urban-rural contrast, in favour of the former. This is true of all the eight variables considered in the analysis. If factor analysed, the variables render a clear general component reflecting housing disamenity. The scores of the various administrative areas making up the three countries on the respective first factors is shown in Figure 6.4. In each, housing disamenity reaches a peak in the rural peripheral administrative areas, and displays negative values in respect to the city centres and the suburban zones.

Housing policies in the region have manifestly failed to tackle the existing problems of provision. In each territory the state has failed to incorporate the spontaneous self-help efforts of the ordinary people into an effective plan for housing improvement (Potter 1990), and thus the intermediate or appropriate technology connotations of the local vernacular architectural system have not been tapped in housing policies. There have been no comprehensive state-aided self-help schemes involving upgrading, site and services or core housing. Nor have the three countries concerned initiated programmes of state provision of rental housing (Figure 6.5), although St Lucia

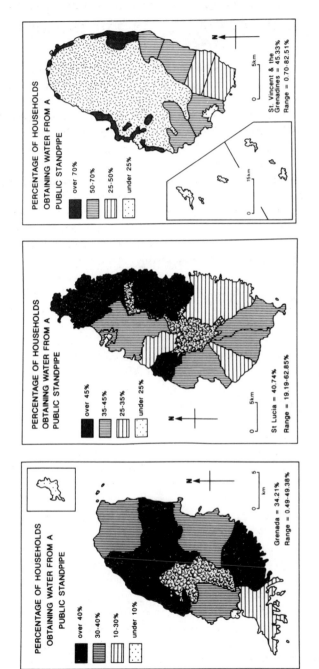

Figure 6.3 Geographical variations in the use of public standpipes in Grenada, St Lucia and St Vincent and the Grenadines

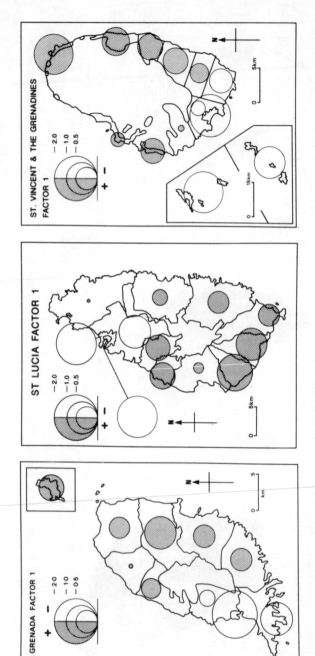

Figure 6.4 The scores of administrative areas on the housing amenity factor derived for Grenada, St Lucia and St Vincent and the Grenadines

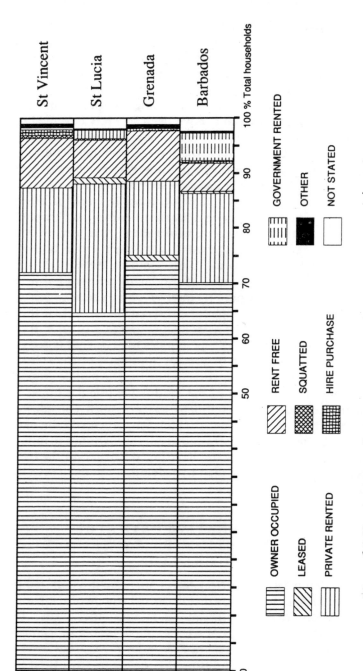

Figure 6.5 Housing tenure characteristics in St Vincent, St Lucia, Grenada and Barbados

in particular, briefly flirted with the idea. Further, none of the three nations has a current housing plan and the respective Ministries of Housing, which are generally amalgamated with other ministries, have little if anything to do with the construction of housing *per se*. In each this function is vested in a technically-oriented National Housing Corporation. The achievements of these units have been severely curtailed by the requirement that they operate on a self-financing basis. They are expected to recycle the returns from one scheme in order to buy the land for the next. Thus, very few houses have been built, and those that have are well beyond the means of the rural and urban poor that are most in need of them. Further, the housing finance opportunities available through the national and commercial banks mean that low-income groups are effectively excluded altogether from housing finance (Phillip 1988; Louis 1986; Ishmael 1989).

In all these ways the onus of producing housing has been squarely placed on the poor themselves, and it can be suggested that these states have sought to conserve the vernacular form as a cheap means of social reproduction (McGee 1979; Burgess 1982). In the few instances where states have ostensibly sought to pursue the dissolution of popular housing, they have ended up producing expensive apartment blocks. The middle ground, involving the utilization of the modifiable popular house in schemes of state-aided self-help has not been considered. Notably, it is the rural areas which have received the least attention. Despite their pressing need. It is hard not to ignore the argument that these states have failed to cater for the basic needs of the poor other than by leaving the poor to provide for themselves.

AN ALTERNATIVE PATH: THE POST-REVOLUTIONARY EXPERIENCE OF GRENADA

On 13 March 1979, the New Jewel Movement (NJM), led by Maurice Bishop, overthrew the dictatorial regime of Eric Gairy. In addition to 'anti-Gairyism' and 'anti-imperialism', the principal themes of the NJM were explicitly those of 'genuine independence' and 'self-reliance'. Brierley (1985a) has reviewed in some detail the condition of Grenada on the eve of the revolution, pointing to many of the conditions outlined in the earlier sections of this chapter. These included the effects of the global economic recession, a chronic trade deficit, strong reliance on grants-in-aid and remittances by nationals living overseas, dependence on high-priced food imports, large areas of idle agricultural land and high unemployment. Against this background writers have described Gairy as an autocratic leader who was plundering the country for his own benefit (Brierley 1985a; Jacobs and Jacobs 1980 and Ambursley 1983).

Following the *coup d'état*, the first overthrow of a Commonwealth Caribbean government, the People's Revolutionary Government (PRG) took power under the Prime Ministership of Maurice Bishop. The development path followed has been documented by Brierley (1985a), who notes that it accorded with the basic human needs philosophy espoused in the NJM's earlier manifestos. The main concerns were to prevent the prices of food, clothing and other essentials rocketing, and to improve overall standards of housing, clothing, education, health and recreation for all sectors of the community. As well as freezing food prices, the PRG announced its intention to depart from Grenada's traditional role as an exporter of cheap agricultural production. Although styled as a socialist revolution, as Brierley (1985a) notes, after the coup approximately 80 per cent of the economy remained in the hands of the private sector, and a tri-sectoral strategy of development, encompassing the private, public and co-operative parts of the economy was advanced.

In the field of agriculture, the thrust toward greater self-sufficiency was closely allied with the programme to 'marry idle lands with idle hands' (Brierley 1984, 1985b, 1985c). This was executed via the National Co-operative Development Agency established in 1980, with groups of unemployed people identifying with land in their neighbourhood and indicating how they would work it as a co-operative venture. At the same time, two agro-industry plants were completed between 1981 and 1982, one producing coffee and spices, the other juices and jams.

The construction of a new international airport at Pt Salines was undertaken during the first year of the revolution (Hudson 1983). The bulk of the US $75 millions expended was met by Cuba, a contributing factor behind the American decision to invade in October 1983. With the new airport came what the PRG promoted as the 'New Tourism'. The principle behind this was the development of what were described as 'sociologically relevant forms of tourism' – those which emphasize cultural themes and which are based on local foods, cuisine, handicrafts and furniture-making.

Other developments in the social services – such as the provision of free medicine, dental and educational services – were just as significant in making efforts to provide for the basic needs of the poor. It seems clear that in these regards, the PRG achieved a great deal in the four and a half years it was in power. The assassination of Bishop, and key members of his Cabinet, in October 1983 and the invasion by the United States which followed brought to a premature end the experiment which had started in 1979. This was a pity, for as Brierley (1985a) notes, it deprived other small less-developed countries of the lessons of grassroots development that would have been provided if the political and socio-economic experiment had run its course. Saliently, during the period from 1979 to 1983, the economy grew quite substantially, at rates between 2.1 and 5.5 per cent per annum, a feature which was commented upon favourably by the World Bank (Brierley, 1985a). During the same period the value of Grenada's imported foodstuffs fell from 33 to 27.5 per cent.

CONCLUSIONS

It has to be fully recognized that small island sates such as those found in the Eastern Caribbean face many difficulties in endeavouring to reduce even marginally their formidable reliance on large metropolitan nations. This is reflected in the fact that, since independence, with the exception of the PRG government of Grenada, no large-scale, comprehensive efforts have been made to provide for the basic needs of the poor majority. Only small-scale and token schemes have occurred, such as those supported by the Basic Needs Fund. This was administered in the Eastern Caribbean by the Caribbean Development Bank in the late 1970s and the early 1980s, via finance derived from the United States Agency for International Development (USAID) (Downes 1980). It provided for basic infrastructural facilities in the fields of education, health, water supply, drainage, forestry and roads, but operated largely on a palliative basis.

It must be reiterated that achieving even a limited degree of selective closure of the economy in small islands presents a series of almost insuperable problems. This chapter has, however, posited that some considerable degree of basic-needs orientation is both possible and highly desirable within the small island states that make up the Eastern Caribbean. Such approaches must stress indigenous requirements and resources, and must constitute a distinct rejection of exogenously-defined ideologies of change and development. Thus, there is pressing need to move toward more sustainable and locally sensitive forms of tourism and industrial development. In the past, states have all too often abdicated their responsibility to help the people in their efforts to help themselves. In so doing, governments have either promoted inappropriate solutions, or have left the disadvantaged to fend entirely for themselves. The classic instance of this is, of course, housing, as exemplified earlier in this chapter. The example of Grenada suggests that the situation is not impossible. Equally, however, the response does not have to be tied to a socialist transformation, but can be effected to some considerable degree within a mixed economy that takes on board co-operative ventures, the imperatives of rural-based development, and the salience of indigenously-defined patterns and processes of socio-economic change.

ACKNOWLEDGEMENT

This work forms part of a project supported by the Economic and Social Research Council on housing and development in the Eastern Caribbean.

REFERENCES

Ambursley, F. (1983) 'Grenada: the New Jewel Revolution', Ch.9 in F. Ambursley and R. Cohen (eds) *Crisis in the Caribbean*, London: Heinemann.

Barry, T., Wood, B. and Preusch, D. (1984) *The Other Side of Paradise: Foreign Control in the Caribbean*, New York: Grove Press.

Beckford, G.L. (1972) *Persistent Poverty: Underdevelopment in Plantation Economies of the Third World*, New York: Oxford University Press.

Brierley, J.S. (1984) 'Idle land and idle hands in Grenada, West Indies', *Rural Systems* 2: 233–44.

Brierley, J.S. (1985a) 'A review of development strategies and programmes of the People's Revolutionary Government in Grenada, 1979–83', *Geographical Journal* 151: 40–42.

Brierley, J.S. (1985b) 'The agricultural strategies and programmes of the People's Revolutionary Government in Grenada, 1979–83, *Conference of Latin American Geographers Yearbook* 55–61.

Burgess, R. (1982) 'Self-help housing advocacy: a curious form of radicalism: a critique of the work of J.F.C. Turner', in P.M. Ward (ed.) *Self-Help Housing: A Critique*, London: Mansell.

Central Bank of Barbados (1989) *Economic and Financial Statistics*, Bridgetown, Barbados: Central Bank of Barbados.

Connell, J. (1987) 'Contemporary issues in island micro-states', Ch.13 in M. Pacione (ed.) *The Geography of the Third World: Progress and Prospects*, London and New York: Routledge.

Downes, A.S. (1980) 'A basic needs strategy for newly independent micro-states', *Bulletin of Eastern Caribbean Affairs* 6: 12–22.

Drakakis-Smith, D. (1991) 'Food for thought or thought for food: urban food distribution systems in the Third World', Ch.7 in R.B. Potter and A.T. Salau (eds) *Cities and Development in the Third World*, London and New York: Mansell.

Fraser, P.D. (1985) *Caribbean Economic Handbook*, London: Euromonitor Publications.

Gordon-Somers, T. (1984) 'New dimensions revisited: a case for the small islands', *Bulletin of Eastern Caribbean Affairs* 10: 18–25.

Gumbs, F. (1981) 'Agriculture in the wider Caribbean', *Ambio* 10: 335–9.

Hall, D.R. (1989) 'Cuba', Ch.4 in R.B. Potter (ed.) *Urbanization, Planning and Development in the Caribbean*, London and New York: Mansell.

Harrison, R. (1984) 'English speaking Caribbean less developed countries (LDCs): growth and development of the manufacturing sector', *Bulletin of Eastern Caribbean Affairs* 9: 1526

Hettne, B. (1990) *Development Theory and the Three Worlds*, Harlow: Longman.

Hudson, B.J. (1983) 'Grenada's new international airport', *Caribbean Geography* 1: 51–7.

International Labour Organization (1976a) *Employment, Growth and Basic Needs: a One-World Problem*, Geneva: ILO Office.

International Labour Organization (1976b) *Declaration of Principles and Programme of Action for a Basic Needs Strategy of Development*, Geneva: ILO Office.

Ishmael, L. (1989) *Housing Sector Overview: Sites and Services and Squatter Upgrading*, Kingstown, St Vincent: Government of St Vincent and the Grenadines.

Jacobs, W.R. and Jacobs, B.I. (1980) *Grenada: the Route to Revolution*, Havana: Casa de las Américas.

Kelly, D. (1986) 'St Lucia's female electronics factory workers: key components in an export-oriented industrialisation strategy', *World Development* 14: 823–38.

Kowalewski, D. (1982) *Transnational Corporations and Caribbean Inequalities*, New York: Praeger.

Lewis, W.A. (1955) *The Theory of Economic Growth*, London: Allen & Unwin.

Lloyd-Evans, S. and Potter, R.B. (1991) 'The informal sector of the economy in the Commonwealth Caribbean: an overview', *Royal Holloway & Bedford New College, Papers in Geography* 8.

Louis, E.L. (1986) *A Critical Analysis of Low Income Housing in St Lucia*, MSc dissertation, St Augustine Campus of the University of the West Indies.

McGee, T. (1979) 'Conservation and dissolution in the Third World city: the "shanty town" as an element of conservation', *Development and Change* 10: 1–22.

Macleod, S. and McGee, T. (1990) 'The last frontier: the emergence of the industrial palate in Hong Kong', Ch. 11 in D. Drakakis-Smith (ed.) *Economic Growth and Urbanization in Developing Areas*, London and New York: Routledge.

Marshall, D. (1982) 'Migration as an agent of change in Caribbean island ecosystems', *International Social Science Journal* 34: 451–67.

Momsen, J.D. (1986) 'Migration and rural development in the Caribbean', *Tijdschrift voor Economische en Sociale Geografie* 77: 50–8.

Nanton, P. (1983) 'The changing pattern of state control in St Vincent and the Grenadines', Ch. 10 in F. Ambursley and R. Cohen (eds) *Crisis in the Caribbean*, London: Heinemann.

O'Loughlin, C. (1968) *Economic and Political Change in the Leeward and Windward Islands*, New Haven and London: Yale University Press.

Phillip, M.P. (1988) *Urban Low Income Housing in St Lucia: An Analysis of the Formal and Informal Sectors*, unpublished MPhil thesis, University of London.

Population Census of the Commonwealth Caribbean (no author) (1980–1) vol. 3, Printing Unit, Statistical Institute of Jamaica.

Potter, R.B. (1980) 'Industrial development and urban planning in Barbados', *Geography* 66: 225–8.

Potter, R.B. (1983) 'Tourism and development: the case of Barbados, West Indies', *Geography*, 68: 46–50.

Potter, R.B. (1989a) 'Rural-urban interaction in Barbados and the Southern Caribbean: patterns and processes of dependent development in small countries', Ch. 9 in R.B. Potter and T. Unwin (eds) *The Geography of Urban-Rural Interaction in Developing Countries*, London and New York: Routledge.

Potter, R.B. (ed.) (1989b) *Urbanization, Planning and Development in the Caribbean*, London and New York: Mansell.

Potter, R.B. (1989c) 'Urban housing in Barbados, West Indies', *Geographical Journal* 155: 81–93.

Potter, R.B. (1990) 'The quality of housing in Grenada, St Lucia and St Vincent: a cartographic and statistical analysis', *Royal Holloway & Bedford College, Papers in Geography* 7.

Potter, R.B. (1991a) 'Cities, convergence, divergence and Third World development', Ch.1 in R.B. Potter and A.T. Salau (eds) *Cities and Development in the Third World*, London and New York, Mansell.

Potter R.B. (1991b) 'A note concerning housing conditions in Grenada, St Lucia and St Vincent', *Bulletin of Eastern Caribbean Affairs* (in press).

Potter, R.B. (1991c) 'An analysis of housing in Grenada, St Lucia and St Vincent and the Grenadines', *Caribbean Geography* 3: 107–25.

Potter, R.B. (1992) *Housing in Barbados: a Geographical Analysis*, Kingston, Jamaica: University of the West Indies.

Potter, R.B. and Dann, G.M.S. (1991) 'Dependent urbanization and retail change in Barbados, West Indies', Ch.11 in R.B. Potter and A.T. Salau (eds) *Cities and Development in the Third World*, London and New York: Mansell.

Rojas, E. (1984) 'Agricultural Land in the Eastern Caribbean: from resources for survival to resources for development', *Land Use Policy* 1: 39–54.

Rojas, E. (1989) 'Human settlement of the Eastern Caribbean: development problems and policy options', *Cities* 6: 243–58.

Rojas, E. and Meganck, R.A. (1987) 'Land distribution and land development in the Eastern Caribbean', *Land Use Policy* 4: 157–67.

Rubenstein, H. (1983) 'Remittances and rural underdevelopment in the English-speaking Caribbean', *Human Organization* 42: 295–306.

Shephard, C.Y. (1947) 'Peasant agriculture in the Leeward and Windward Islands', *Tropical Agriculture* 24: 61–71.

Thomson, R. (1987) *Green Gold: Bananas and Dependency in the Eastern Caribbean*, London: Latin American Bureau.

World Bank (1988) *The World Bank Atlas*, Washington DC: The World Bank.

7

ISLAND MICROSTATES

Development, autonomy and the ties that bind

John Connell

There are twenty politically-independent island microstates in the developing world – that is states with populations of less than a million (Table 7.1). This chapter seeks to examine development trends in these states, to reflect on the meaning of independence, and to compare and contrast the experience of these states with dependent and often neighbouring territories and colonies. It particularly focuses on the extent to which the structure of economic change is conducive to long-term development. Definitions of development have been legion, mainly revolving around issues such as basic needs, equity, self-reliance and power. More than twenty years ago the economist Dudley Seers suggested, 'The questions to ask about a country's development are What has been happening to poverty? What has been happening to unemployment? What has been happening to inequality?' (Seers 1969:3). This perspective subsequently expanded into the concept of 'redistribution and growth' (Chenery *et al.* 1974). Seers expanded his perspective to embrace 'greater independence with redistribution and growth' (Seers 1977: 5) and much debate surrounded the definition and utilization of various concepts of basic needs (Connell 1980) few of which concluded anything other than that development was complex, multifaceted and relative. Wallman concluded that development was 'a progression towards the (better?) meeting of basic needs, and it is a progression towards (greater?) autonomy and authenticity of self and/or nationhood' (Wallman 1977:12). It is in the context of these diffuse perspectives on development that this chapter is situated.

Table 7.1 Island microstates

	Population (mid-1988)	Area (sq. km)	Population density (persons per sq. km)	GNP per capita (US $ 1988)	Life expectancy (1988)
Caribbean					
Antigua and Barbuda	78,000	442	176	3,690	73
Bahamas	244,000	13,940	18	10,700	68
Barbados	254,000	430	59	6,010	75
Dominica	82,000	750	109	1,680	74
Grenada	94,000	345	272	1,720	69
St Kitts-Nevis	42,000	269	156	2,630	69
St Lucia	145,000	616	235	1,540	71
St Vincent and the Grenadines	12,000	388	288	1,200	70
Indian Ocean					
Comoros	442,000	1,865	236	440	56
Maldives	202,000	298	678	410	60
Seychelles	68,000	404	168	2,800	70
Atlantic Ocean					
Cape Verde	360,000	4,033	89	680	65
São Tomé and Principe	119,000	826	144	490	65
South Pacific					
Fiji	732,000	18,272	40	1,520	71
Kiribati	67,000	690	97	650	55
Solomon Islands	303,000	28,530	11	630	64
Tonga	97,000	700	139	830	66
Tuvalu	9,000	26	346	570 (1980)	68
Western Samoa	159,000	2,935	54	640	66
Vanuatu	147,000	11,880	12	840	64

Sources: World Bank 1990: 243. The data for Tuvalu are from the *Pacific Economic Bulletin* 1990, 5(1): 36. None of these data can be considered to be wholly accurate, hence the table is included solely for comparative purposes.

Most island mirostates became independent in the 1960s and 1970s, an era of extraordinary optimism over the prospects of development in the emerging states of the Third World. Even so, not all tiny states sought independence willingly; some had independence largely thrust upon them (for example Kiribati, Solomon Islands) whilst many small islands (notably Anguilla) struggled successfully against any prospect of independence. Few (like Vanuatu) were forced to fight for independence. Newly-emerging states were offered 'a model of development to avoid' (Hart 1974) – based on the disappointing experiences of sub-Saharan African states – on the assumption that the principle gain from 'late development', or at least late independence, was that the successes and failures of previously independent states would provide powerful lessons. Development strategies, one way or another, sought to build on what had gone before.

The development plans and policies of the newly-independent states were oriented to the classical task of maximizing local resources, though such were recognized to be often extremely limited. They focused on the export of agricultural and fisheries products (and minerals in the rare cases where these existed), assumed that industrialization would eventually follow, and that this would be oriented initially to import-substitution and the processing of agricultural goods, along the lines of Michael Lipton's dictum 'if you wish for industrialisation prepare to develop agriculture' (Lipton 1977:24). In this scenario the role of tourism was ambivalent; in the Caribbean it was already well-established, but in the Pacific it was less welcome because of its feared cultural implications and hence seen, in every way, as something of a 'last resort' (Lea 1980). Development strategies were thus evolutionary rather than revolutionary, stressing the necessity for a greater degree of self-reliance, which would inevitably follow the emphasis on agriculture (especially subsistence agriculture) and fisheries. Even the most dependent territories sought greater self-reliance – by which time the basic philosophy had become little more than a rhetorical incantation. Eventually, however, it was recognized that, as the former President of Palau observed, 'we have to educate the people to the need for sacrifice. So we will have to use dependency to achieve self-sufficiency' (quoted in Connell 1988a:16). Aid was therefore widely expected to play some role in the structural transformations required to achieve greater self-reliance.

Clearly it was not going to be easy for island states to achieve either greater self-reliance, or, in any conventional sense, the transition to developed country status. Such concerns had often delayed independence, as both colonial administrators and dependent populations feared that small size might be a constraint to development. A litany of now familiar problems emphasized the difficulties of development: remoteness and isolation (hence high transport costs to remote markets), diseconomies of scale, limited natural resources, substantial trade deficits (but considerable dependence on trade with metropolitan states), few local skills, vulnerability to hazard, disproportionately high expenditure on administration, political fragmentation and a dependence on external institutions for some key services (for example universities and banks). Political systems were fragile, ecological structures were vulnerable and economies lacked diversity. Set against these disadvantages were few obvious advantages: isolation had led to some degree of cultural preservation (but language groups were often so small that their integrity was often disrupted by a modern world of education and mass media) whilst isolation had led to many remote islands having strategic value, a situation that often enabled financial gains, but not without considerable social cost. On balance, the constraints to development appeared to overwhelm the advantages, a situation that was dismissed in the immediate post-colonial era as a first generation of leaders undertook to transform new nations. The stage was set.

THE DEVELOPMENT OF NATURAL RESOURCES

Island microstates (IMS) were characterized by their agricultural economies, a situation which usually continued well into the second half of the twentieth century. Outside the Pacific, plantation economies historically dominated the agricultural sector, but in the twentieth century the plantation system, and the 'coconut overlay' reached the Pacific islands, freezing land areas, and ushering in a world of monetization and commodity consumption. The inheritance of plantation systems has often proved a burden rather than a blessing, land reform has become a constant theme of agricultural development, especially in the context of (usually) rising population numbers and virtually nowhere does agriculture still characterize the economies of island microstates, except in terms of employment (though not in the formal sector).

A small numer of IMS – notably the two sub-Saharan states Cape Verde, and São Tomé and Príncipe – and the Indian Ocean state of the Comoros – have experienced agricultural problems that have brought them close to famine. Their situation and, to a lesser extent, that of such atoll states as the Maldives and Kiribati, are of continued concern. There and elsewhere a number of general trends have influenced the direction of agricultural development. First, attitudes towards agriculture and agricultural employment have substantially changed; in parallel with the increased need to earn cash, agricultural work throughout the IMS has lost prestige and the declining participation of young men especially is ubiquitous. In Fiji youths 'are taught to value white-collar occupations and farming' (Naidu 1981:8). Such attitudes, alongside the expansion of urban employment opportunities (at home or abroad) have led to migration (and emigration), increasing the dependency ratio in many rural areas, leading to an ageing and often more female population, imposing heavy burdens on those who remain, hence often leading to declining agricultural production both in total and per capita.

Changing attitudes to agricultural employment parallel changing attitudes to food consumption. Changing tastes, monetization, prestige and the cost and availability of imported foods have led to a rapid movement towards 'food dependency', a situation where food and beverage imports are disproportionately high. The simultaneous decline in subsistence agricultural production and growing consumption of imported foods (especially tinned meat and fish, rice, biscuits and flour) has both reduced the regularity and variety of food consumption (as cash flows are variable) and substantially increased the incidence of diet-related diseases such as diabetes. There has been a distintensification of traditional agricultural systems (and a switch to low intensive crops such as cassava, with negative nutritional consequences), a loss of variety (in gardens and through hunting, gathering and fishing), and the resultant loss of what has been termed in Fiji the 'subsistence safety net' (Taylor 1987).

Agricultural economies, even in historical plantation states, have usually become increasingly specialized, favouring a small number of export crops, such as sugar, and these crops have generally benefited from the concentration of public expenditure (research funds, loans, infrastructure provision, and so on) in these export-oriented areas. By contrast public expenditure on food crops is notoriously lacking. Although at particular times IMS have achieved niches in the world market economy for such minority crops as ylang-ylang (Comoros), ginger (Fiji) and vanilla (Tonga), these have only exceptionally proved to be long-lasting and capable of supporting more than a minority of farmers. More generally, even relatively large states such as Mauritius are highly dependent on a single export crop. In exceptional circumstances, however, some IMS have filled a new niche with one particularly successful product – marijuana. Even in the large island state of Jamaica, it is the single most important export by value (though no statistics record this fact); in Palau it is also the most valuable export. In St Vincent it has not only brought new wealth into the community but 'it is fostering a reduction in the stigma attached to working land amongst young people, and this may be its most important legacy' (Rubenstein 1988:130). It is, of course, universally illegal and poses certain obvious problems.

There are several disadvantages to the concentration upon and the export of a small range of crops from IMS, including ecological problems, the similarity with other IMS (and with much larger tropical countries, where there are economies of scale and better market access) and the dependence on world market price fluctuations. What has been crucial to success has been the ability of many IMS to negotiate bilateral agreements that give special advantages and thus avoid competing at global price levels; exclusion from free market conditions has further emphasized the concentration on export crops. This has been particularly true for sugar producers, with preferential access to both the European Economic Community (EEC), through the Lomé Convention, and the USA. Fiji, for example, sells three-quarters of its sugar under favourably-priced contracts giving it 'a degree of price stability and security and . . . in recent years, prices above the world market average'

(Fairbairn 1985:99). Thus, where special agreements have been reached, as in the Caribbean, 'the misery of growing cane is nothing compared to the misery of growing other crops' (World Bank 1987:3). Away from particularly-favoured commodities, such as sugar, the prospects are more limited; copra producers, for example, have few advantages. In a sense 'this process of subsidy amounts to a type of annually renewable aid package. But this is no ordinary aid package. Instead it is an aid package operated and administered by the full machinery of a national economy. This is aid with dignity' (Taylor 1987:3). Thus Fiji, like other IMS, has no genuine comparative advantage in sugar production (as some assumptions about efficiency suggest) but, in fact, has 'a comparative advantage in the garnering of subsidies and long-term and short-term contracts for whatever it produces' (ibid.:10). Protected markets and subsidized prices have been crucial to agricultural development in IMS. On balance, the high and stable prices paid for crops that are currently grown in the IMS appear to have overcome anxieties about longer-term agricultural diversification. Moreover, there is some probability that preferential access for most IMS will continue (Brookfield 1987:56–7). For the moment those IMS that are able to produce appropriate agricultural commodities, and gain access to preferential prices, can assume that these prices will not decline dramatically.

The disadvantages and limited prospects of agricultural development beyond 'protected' crops have convinced many observers that a continued focus on increased agricultural production in national development strategies for IMS may be unwise. Thus a review of South Pacific trade concluded that 'the expansion of manufacturing capacity should be encouraged to offset diminishing returns to land. In this effort the guide should probably be the successful experience of East Asian countries which had adapted outward-looking, export-oriented industrialisation rather than import-substitution industrialisation' (Charle 1986:30). How this might be achieved was unstated and, so far, industrialization of any kind, especially in the Pacific IMS, has been limited in its achievements.

Broadly the same kind of cumulative downward spiral that has affected agriculture has also affected artisanal fisheries in IMS, though less dramatically and less well documented. Considering the great extent of the fishing grounds that surround IMS, and the often limited role of agriculture, artisanal fishing is an extremely underdeveloped economic activity. Several common problems have contributed to this: a lack of biomass (especially in atoll states), overexploitation and the costly and uneconomic purchase of boats, gear and fuel. Fishermen are too often viewed, much like farmers, as unworthy of serious government investment of energy and resources. In Mauritius, for example, there is reported to be an 'island-wide mentality to view traditional fishermen with undignified contempt and to blame their distress and misery on insufficient volition and salubrity' (Paul 1987:144). Marketing infrastructure for an extremely perishable product is often inadequate and, as in the Caribbean, 'the policy of most governments has appeared to be in token form' (Walters 1984:95), even behind that for agriculture. Successes have been few and far between.

Exploitation of the fisheries potential of the vast areas within the 200-mile exclusive economic zones (EEZ) around IMS is an even more challenging prospect. Current development potential is primarily restricted to the leasing of these areas to the deep-water fishing vessels of distant, richer nations. International restructuring, incorporating more capital-intensive purse-seiners demanding sophisticated technical skills, at a time of global market saturation, and lack of onshore facilities, have restricted the attempts of IMS to participate in this sector. Moreover the EEC STABEX (export earning stabilization) scheme, which provides concessionary finance for most agricultural exports, does not apply to exports of fish. Thus IMS have been primarily dependent on free market prices for fisheries development and this may constitute a further reason for the limited development of national fisheries in IMS.

Only three IMS, Fiji, Mauritius and the Solomon Islands, have developed fisheries to the extent that they now operate nationally-owned or joint-venture canning factories. IMS have yet to gain significantly from the new EEZ legislation: moreover, the countries of the Eastern Caribbean have acquired exclusive rights 'to some of the most biologically unproductive waters in the region' (Dolman 1985:58). In the Indian and Pacific Oceans the potential is greater, yet such gains, because of non-existent technical capacity to exploit them, are more theoretical than real, and even policing these waters is extremely difficult. The importance of Pacific IMS has been apparent in their almost total inability to restrict illegal fishing, and for the foreseeable future the likelihood of developing internationally-competitive national fishing industries is very poor. It is, however, one of the few realistic economic development prospects for this group of IMS and perhaps for some in the central Indian Ocean.

Most fisheries processing is undertaken in rich countries, or territories such as American Samoa. Fish prices are as flexible as cash crop prices; the collapse of export prices, and the withdrawal of Japanese fishing companies in 1982–3, led to the collapse of the promising Maldives fishing industry (Sathiendrakumar and Tisdell 1986:281), though in the Pacific a similar collapse enabled Pacific IMS, through the South Pacific Forum Fisheries Agency, to eventually regulate the depredatory fishing activities of vessels of the American Tunaboat Association and achieve improved access fees. However, this change followed both a decline in the value of fish imports into the USA and an upward revision by the United States of the strategic importance of Pacific IMS, which partly followed the leasing of fishing rights by Kiribati and Vanuatu to the Soviet Union. Most IMS have entered into some form of agreement with metropolitan states, or with the EEC, for the leasing of national waters. IMS, who do gain much income from fisheries, do it primarily through leasing their waters rather than through owning fishing fleets. However, the IMS countries receive less than 5 per cent of the value of catches in their waters in licences fees (Fairbairn 1985:82). Though such returns are small, they constitute a major source of revenue for some IMS.

Few IMS are minerals producers. Fiji is, however, a significant gold producer and has gained from the gold price rises of the 1980s. Nearby Solomon Islands and Vanuatu, astride the Pacific 'rim of fire', are also poised to establish mining industries, though by their nature, such industries are foreign-owned, generate little employment, may contribute to a 'Dutch disease' and pose some ecological problems. There are also prospects for the marine mining industry, but none appear imminent, whilst energy – in growing quantities – must also invariably be imported.

The development of the natural resources (to which the timber industry might be added) of IMS has thus become increasingly tied to the world political and economic system. Subsistence activities, whether in agriculture or fishing, have steadily given way to commodity production. Despite official emphasis on diversity, monoculture has thrived, and productive activities that are rural, labour-intensive, involve female participation and incorporate appropriate (non-imported) technology have been replaced by their converse. Despite subsidized agriculture, and the rental of coastal waters, national incomes derived solely from natural resources have been disappointing, and ensured that IMS have sought to diversify beyond their supposed comparative advantage – the agricultural and fisheries economy – into other arenas.

INDUSTRY, TOURISM AND BEYOND

Two such sectors – manufacturing industry and tourism – are quite conventional and, in the latter case, were well established in some IMS prior to independence. Classical industrialization, the establishment of import-substitution industries, has proved difficult because of standard constraints such as the small size, fragmentation and low earnings of the domestic market, limited skills, few raw materials, inadequate access to technology and investment capital, high energy costs and the lack of tariff protection for new industries. Most such industries are agro-food industries (bakeries, breweries, and so on) characterized, in the Eastern Caribbean, as being the 'Coke, Curtains, Coconut Cream and Corn Curls' phase of industrial development (G. Theophilus, quoted in Connell 1988a:53).

Historically, the establishment of export-oriented industries proved even more difficult. In addition to most of the problems that faced import-substitution industries, were extra problems: access to markets (in terms of both infrastructure and tarrifs), small volumes of goods and high labour costs. In conditions of free trade the IMS had few competitive manufactured products. Open economies, enabling relatively cheap, untaxed imports have discouraged domestic investment (and much foreign investment). Where industrialization has been successful it has either, as in the case of the Seychelles, been hidden behind tariff barriers or it has been highly dependent on the combination of private foreign investment and preferential access to metropolitan markets. In manufacturing especially, 'the foreign sector is the economy'. (Dommen and Hein 1985:152). As the United Nations Conference on Trade and Development (UNCTAD) recorded in 1985: 'in a small country with no free trade, industrial growth may be inefficient, however, with free trade a small country may experience no growth at all and may in fact deindustrialise' (UNCTAD 1985:25).

Dependent industrialization or 'industrialisation by invitation' (Barry, Wood and Preusch 1984:73) has been highly successful in a number of states. Spearheaded by low-wage Haiti, and assisted by the tariff exemptions of the 1982 Caribbean Basin Initiative, the growth of manufacturing is most true of Caribbean IMS, where high technology 'screwdriver' industries in Export Processing Zones (EPZs) are increasingly common. The factors that have contributed to industrialization there, and in other IMS, include low wages and weak unions, substantial tax concessions, the 'freedom to repatriate capital and profits' (as advertised in post-coup attempts to encourage industrialization in Fiji) and preferential access to the markets of metropolitan states. Caribbean IMS for example, have preferential access to the USA (under the Caribbean Basin Initiative – CBI), to Canada and to the EEC (under the Lomé Convention); Pacific IMS have access to Australia and New Zealand, through the South Pacific Area Regional Trade and Economic Co-operation Agreement (SPARTECA) and also to the EEC. Though Brookfield has queried: 'what greater form of dependence is there than imitation (Brookfield 1975:202), industrialization now generates significant exports and incomes in several states. Despite new employment opportunities there have, however, been costs; women, and most workers in the new EPZs are women, work in repressive, authoritarian conditions for very low wages. Many see factory employment as a means of acquiring industrial skills for subsequent emigration. Governments are rarely capable of reducing the substantial profit transfers of transnational corporations (TNCs), hence increased employment is often the principal and only gain. Characteristic of the 'new manufacturing' is the production of goods, whose unfamiliar raw materials must be imported and which have limited utility in the IMS; woollen garments are the major manufactured exports from Mauritius and Tonga, though few of the women workers have ever seen a sheep or wear such garments. Production is wholly dependent on external market conditions. Not surprisingly Mauritius has been referred to as a 'pyjama republic' (Jones 1989).

With few exceptions there would be virtually no significant export industries without considerable positive discriminatory legislation in their favour, both in the IMS and in metropolitan states. Though all trading agreements, such as the Lomé

Convention and SPARTECA, impose substantial restrictions on the range and volume of imports (that are now causing considerable frustration in Mauritius and Fiji), they have been crucial to industrialization and thus to economic diversification. Privileged access is always under threat in the destinations, hence 'the need for continuous political negotiation at the international level – an expensive proposition for a small country' (Hein 1986:15). Clearly privileged access depends on the political and economic situation in metropolitan states. One view of the CBI is that, 'it has appeared at a time when these struggling countries have been forced to abandon their conventional inward-looking approaches to development, by which they had tried unsuccessfully to insulate themselves from world market pressures. Instead, they have begun to adopt export-promoting strategies similar to those that have proved so successful in East Asia (Conklin 1987). Ironically, yet significantly, this particular article is sub-titled 'a regional solution for America's threatened enterprise?'. In many respects, even relatively successful industrial development in IMS has been primarily dependent on the policies of both metropolitan governments and manufacturers.

As elsewhere in the world, the service sector has increasingly become the most dynamic sector in IMS. Historically characterized by tourism, the service sector has now expanded to incorporate data-processing industries (especially in the Caribbean) and tax havens, as the finance industry has decentralized from heavily-taxed metropolitan states. In a number of countries tourism is the single most important contributor to the gross domestic product (GDP). The few IMS where tourism is unimportant – the Atlantic IMS, the Comoros and the central Pacific – are amongst the poorest of the less-developed countries (LDCs); international transport constraints, the costs of inaccessibility and limited facilities have prevented the genesis of the industry. In Tonga, where tourism is little more developed, there are explanations beyond concerns over harmful cultural influences:

> without the involvement of foreign commercial and political interests, Tonga has not evolved the essential ties with metropolitan markets and their tourism companies. It would seem that Tonga's tourist industry has

paradoxically suffered because the country was not exploited as a fully-fledged colony.

(Britton 1987: 131)

This is no paradox. Access to metropolitan states is critical since there is minimal indigenous tourism.

Tourism constitutes perhaps the only economic sector where there are genuine comparative advantages for IMS: clean beaches, unpolluted waters, warm weather and at least the vestiges of distinctive cultures, though sometimes these turn out to be illusory. Competition between countries is considerable, tourism is unusually subject to the whims of fashion and wages are often low. Structurally it is in some respects analogous to industrial EPZs, though its concessional status is less evident; it is vulnerable to economic uncertainty (domestic and international), local and regional political instability and to ecological changes that often result from tourism itself. The tourist industry has been advised to abandon its vestigial historic concern over cultural change. The World Bank, noting that the future expansion of tourism in Fiji may be limited, because of inadequate superior quality shopping, hotel and leisure facilities and sightseeing, also emphasized that Fiji was deficient since 'the local culture has not been tailored and projected so as to attract tourist attention as has been done in Hawaii and Bali' (World Bank 1987:14). None the less the most successful tourist states (such as Barbados, Bahamas and also Bermuda) enjoy the highest living standards and lowest unemployment rates of all IMS. As Momsen has concluded for Caribbean IMS:

> In the present climate of international relationships tourism holds especially good prospects for the Caribbean as compared to other industries because it constitutes trade with the wealthiest countries in the world in a situation in which protectionism, which in this case would be the restriction of travel by North Americans and Europeans, is far more difficult to impose than it is to impose on visible exports from Third World countries.
>
> (Momsen 1986: 23)

Though this is undoubtedly true of the more accessible Caribbean, and has been true for some IMS elsewhere, it is apparent that it will be difficult for such gains to be realized in the same manner in the more remote Indian, South Pacific and Atlantic Ocean IMS, or for there to be much localization of the industry.

In the search for diversity, financial service industries have gained in their attractions to the IMS, as other sources of earning have dwindled, and have certain advantages for IMS:

> precisely because their economies lack internal linkages, there is little difficulty in designing a set of tax advantages which not only do not weaken the domestic tax base but actually widen it beyond what the local economy itself could achieve.
>
> (Dommen and Hein 1985: 166)

For overseas corporations they enable international profit shifting through transfer pricing and hence reduction of domestic tax liabilities. IMS newcomers, forced to compete with established players, have sometimes found the going difficult; problems of distance, competition and inappropriate infrastructure have restricted the growth of even this extremely flexible and mobile activity. Its future growth is wholly dependent on the taxation regimes of metropolitan states though, where secrecy provisions are considerable, as in Vanuatu, tax havens are likely to withstand many changes in metropolitan legislation.

THE FIRST MIRAGE

The manner in which an economic activity so apparently obscure as the production of postage stamps has become significant to the future of the smallest IMS (Connell 1988c) emphasizes the considerable difficulties of achieving economic growth through conventional strategies and, conversely, the degree of success that some countries have had with apparently 'unconventional strategies' – activities such as tax havens or even tourism – that are of minimal significance in larger states. However, the fact that several island states are defined by the UN as Least Developed Countries (LDCs) indicates that disappointments are more widespread than successes. Disappointments are exacerbated by rising populations; with few exceptions, populations of IMS are now as large as they have ever been and growth rates, despite emigration and recent declines in fertility, often remain high. A small number of IMS are only just beginning to experience the demographic transition.

Along with changes in fertility and mortality there have also been changes in the spatial distribution of IMS populations, It is no longer possible to regard islanders as solely rural people. In most cases urbanization has been the result of the rapid post-war and post-independence expansion in government activity and spending, the consequent boom in well-paid, secure, bureaucratic job opportunities, primarily for the educated elite and skilled workers, and the resultant growth in other areas of service employment. Uneven development and urban bias are typical of even the smallest IMS.

Virtually without exception, the IMS have become characterized by emigration and, in this context, as in so many others, the Pacific states have followed the experience of their Caribbean predecessors. As the history of migration demonstrates, its structure is affected more by conditions in the receiving than in the sending countries, and hence is more likely to be controlled at the destination. Despite such controls, which have certainly significantly slowed emigration, the number of islanders overseas is substantial and, in many circumstances, the balance continues to shift away from the IMS; for some IMS – but particularly territories such as the Cook Islands or Monserrat – there are more islanders overseas than at 'home'.

Emigration has two positive advantages which together discourage national opposition; it acts as a 'safety-valve', to reduce population pressure on scarce resources, and it leads to a flow of remittances. In many small states remittances are the principal source of national income; only where tourism is triumphant or where emigration is rare (Melanesia) is this not true. For Cape Verde 'it is no exaggeration to state that this ensures the survival of the country' (Lesourd and Reaud-Thomas 1987:117) and, in Western Samoa, as elsewhere, migrants have become 'the most valuable export' (Shankman 1976:28). In this last case emigration has become so much more the norm that when external constraints prevented or discouraged it, the 'stolen dreams' of young men led to an increase in the suicide rate, indeed an increase approaching epidemic proportions (Macpherson 1990). Opportunities for migration are highly valued by islanders, freedom of

movement is given high priority, a 'transnational corporation of kin' (Bertram and Watters 1985:499) has emerged,
allowing kin groups to colonize and exploit economic
opportunities across a wide range of environments and,
in certain circumstances, fertility may have risen to ensure
greater access to migration opportunities (Connell 1988a:29).
Though migration is not without costs (in terms of higher
dependency ratios, agricultural decline, cultural change and
greater individualism), most countries now seek better overseas
outlets rather than seeking to limit emigration. Cape Verde,
with opportunities constrained in the USA, is now seeking to
diversify its destinations; in 1987 the President declared that
'we think that if we get organised, we will be able to go on
counting on emigration as a major component of our drive
to regulate our economy' (quoted in Connell 1988a:31).
Indeed, where emigration from IMS is difficult, attitudes
to those once forced to emigrate have changed; in Kiribati:

> In earlier days they were called the land-hungry people;
> they were the unfortunate ones who did not have suffi-
> cient land. Now our values have changed. Settling
> overseas, beyond the oceans of our islands, is something
> to be sought after. Why? Because our population is still
> growing. So now many consider them, the resettled ones,
> as the fortunate ones and they consider us to be the
> unfortunate ones.

(Schutz and Tenten 1979:127)

Crucial to the role of emigration for IMS is its future viability,
which necessarily depends on the economic situation and
political decisions in current and potential host nations. Several
recent studies of Pacific IMS have concluded, often reluctantly, that improved migration opportunities for the people
of IMS constitute a genuine form of development assistance.
For example, the Australian aid agency AIDAB has suggested
that 'for those countries with very poor prospects for
self-sustaining development and poor standards of living,
opening up of migration policy may be an essential adjunct
to aid' (AIDAB 1987:34); few islanders would differ with this
conclusion and emigration will continue to represent a
formidable aspiration within the IMS.

The final source of income in IMS is foreign aid; despite the rhetoric of self-reliance, there has been little or no opposition to the principle of sustained and increased aid delivery. Overall per capita aid flows to IMS are exceptionally high by global standards; the IMS have long benefited from a widespread 'small-country bias' in aid delivery. States that have remained in some form of dependent relationship (such as Niue, Pitcairn, the Micronesian states or the French overseas departments and territories – the DOM-TOMs) have been even more fortunate (Aldrich and Connell 1991; Connell 1991). Though IMS have an obvious need for aid, it is much more visible there and is far beyond the economic imperative that poverty demands; 'small states carry the same voting weight in the United Nations as large states, so the strategic and political imperatives of aid tend to favour small countries' (Jackson Report 1984:42). The same review of Australian aid delivery also noted that 'the faster development takes place the better Australian strategic and economic interests will be served. As development programs succeed, the need for aid will decline and ultimately disappear' (ibid:23). Yet aid has disappeared only in exceptional circumstances, where metropolitan countries have exacted punishment for the folly of political decisions taken in IMS (such as immediately, but only briefly, after the Fijian coups). More generally where large regions, such as the Pacific, have been perceived to have a growing strategic significance, traditional donors have increased their aid and new donors have entered the field. Exceptionally these trends have been resisted, but rarely by governments:

> Over half our annual budget is from foreign aid (direct and indirect). Like most other Pacific countries, we've become a permanent welfare case. I can't see us ever getting out of the hole. Many of our leaders don't want to: foreign aid is now built into their view of development, into their way of life. It is also in the interests of foreign powers (our supposed benefactors) to keep us hooked on their aid.
>
> (Wendt 1987:15)

But such views have not found official favour with either donors or recipients and aid levels are likely to remain high, not because aid will utimately contribute to economic growth but because of its political and strategic significance.

The decline or stagnation of the productive sector (especially agriculture) and the growth of imports, offset by aid, remittances and tourist revenue, in a situation where much employment is concentrated in the public sector, has led to the conceptualization of the smallest Pacific microstates as MIRAB economies, dependent on migration, remittances and aid, thus sustaining the burgeoning bureaucracy (Bertram and Watters 1985). The urban bias of MIRAB economies, in aid delivery, bank loans and urbanization (especially of the bureaucracy) and the demise of agriculture and fisheries, suggests that a better acronym would be MURAB (Munroe 1990). Brookfield has gone beyond this to suggest that, since it is government employment that predominates in the bureaucracy, the microstate economy might best be conceived of as a mirage, that is scarcely a genuine economy or a sustainable economy (quoted in Bertram and Watters 1985:497). Thus one of the problems of MIRAB, Bertram, has argued that in such small IMS the thrust of most development planning, with its focus on production, is misplaced since 'aid, philately and migrant remittances are not merely supplements to local incomes, they are the foundations of the modern economy' (Bertram 1986:809). Moreover, they have virtually no negative ecological implications. This analysis, first propounded for Kiribati, Tuvalu and the dependent territories of the Cook Islands, Niue and Tokelau, is applicable, to a greater or lesser extent, to most IMS. The IMS, for better or worse, have overturned the classical theories of economic development as they move into a 'post-industrial' era, without ever having experienced significant industrialization.

THE END OF THE ERA OF DECOLONIZATION

In the smallest IMS and particularly the remaining dependent territories, where aid (and remittances) have been substantial, the structure of the economy has been transformed from subsistence towards subsidy. This has been most apparent in the dependent MIRAB territories, such as Tokelau and the Cook Islands and the emerging Micronesian states, but was perhaps first recorded in the Torres Strait islands (see, for example,

Beckett 1987). In each of these cases the subsistence (and minor export) economy was rapidy eroded. The smallest economies and IMS, by choice, and larger IMS, for want of a superior option, have increased their ties with metropolitan powers and moved from productive towards rentier status. In every case communities do not wish to withdraw towards subsistence and self-reliance, migration of the young has reduced genuine local development opportunities, and metropolitan governments have been willing, reluctantly or not, to construct a new form of bureaucratic dependence.

It is no accident that the initial formulation of the MIRAB concept incorporated both politically-independent island states and dependent territories. Political independence was not a significant variable. In Martinique (Miles 1986:158) and elsewhere the central economic problem is to preserve and enhance the status of rentier economies. Consequently, practices once regarded in a largely negative light, such as tourism and emigration, have widely become enshrined as policies after conventional development strategies have proved disappointing. Industrialization is by imitation and invitation; tax havens are created and territorial waters leased out as dependence is increasingly negotiated. Rising expectations, ecological degradation on land and sea and population growth reduce historically-valid economic options. For the moment rentier economies have been able to diversify into new arenas as governments contemplate a variety of options and, where territories (especially island territories) occupy strategic locations, real prospects for maintaining and enhancing rentier status are very substantial.

Overseas dependent territories are no longer the classic colonies, that once generated a wealth of literature on the evils of unequal exchange, colonialism, dependency, exploitation and uneven development. By contrast they are the recipients of considerable largesse from the centre, most apparent for the smallest territories where exports are minimal and dwarfed by imports (though significant income is also generated through the invisible earnings of tourism or tax havens). Partly in consequence, demands for greater incorporation into the centre have become stronger than pressures for independence. Indeed, it has been argued that departmentalization represents a form of decolonization without independence; Albert

Ramassamy, the Senator for the French Indian Ocean department Reúnion, has suggested that 'for the old colonies that have become departments integration is a form of decolonisation just as much as independence for those who have chosen that' (Ramassamy 1987:8).

In most overseas territories political incorporation has led to the construction of a welfare state (with diverse financial advantages) and, except in the British colonies, has ensured that the ability to migrate to the metropole is a right that is zealously guarded. In the Cook Islands and American Samoa, rights to migrate to New Zealand and the United States respectively have been critical factors discouraging demands for political independence. For the Micronesian states – the Marshall Islands and the Federated States of Micronesia – the movement in 1986 from trusteeship (under the American-administered Trust Territory of the Pacific Islands) to a looser relationship with the United States was accompanied by a formal Compact of Free Association in which the two states sought the provision that any citizens 'may enter into, lawfully engage in occupations and establish residence as a non-immigrant in the United States and its territories and possessions', to ensure that a 'safety-valve' was put in place. Since then there has been considerable migration to Guam and the Northern Marianas, and the start of onward migration to the United States (Hezel and Levin 1990).

With the principle exception of New Caledonia, independence movements in small island territories are absent, and, in a number of the DOM-TOMS, even the tiny independence movements have noted that, though independence would provide the psychological boost of political autonomy, it could lead to some decline in the physical quality of life. Not surprisingly, ideological austerity has usually been combined with minimal support. More generally, as in Bermuda and, to a lesser extent, the Turks and Caicos Islands, there is a generalized concern that factional politics in a small island state would be disruptive and that dependent political status is preferable, as it appears to guarantee the continued success of business activity (especially tourism and tax havens), political stability and security (Connell 1987; 1990). Underlying all debates on changing political and economic status, and hence relationships with the metropole (and with other regional and metropolitan states), are two conflicting issues, well summarized in the case of Guam:

> There is in Guam's quest for political identity a fundamental contradiction in what Guam is trying to accomplish. The Chamorro activists belatedly seized upon self-determination as the major principle behind commonwealth. But self-determination marches under the flag of freedom, whereas commonwealth marches under the banner of equality. Although they may seem to go arm in arm, Alexis de Tocqueville noted long ago that freedom and equality will always be at odds with each other.
>
> (Rogers 1988:25–6)

Political integration, as in the French departments, provides no hope of more self-reliant economic development, or the recognition of local cultural issues and rights. Movement towards more self-reliant economic and political development reduces external financial support and causes local concern over both the quality of life and security.

Independent island states have not surprisingly rarely sought any diminution of political autonomy. However, in Dominica, situated between the French departments of Martinique and Guadeloupe, and in Guyana, so close to the French department of Guyane (French Guiana) there has been intermittent (and almost certainly minority) interest in becoming a French overseas department, (see, for example, McDonald 1989), whilst in the much larger state of Jamaica, the Prime Minister, Michael Manley, has recently observed: 'In the Caribbean we are accelerating the integration process because we will not survive as a set of disparate mini-states, unless we want to become a department of France' (quoted in Gauhar 1989:11). More generally, in IMS political autonomy has rarely met the economic aspirations of islanders. There has been significant (often illegal) migration into Caribbean departments, notably from Haiti to Guadeloupe, from Dominica and St Lucia to Martinique and from Brazil and Surinam to French Guiana. In the Indian Ocean there has been a massive recent migration from the Comoros to Mayotte. Other Caribbean island territories, notably Puerto Rico and the American and British Virgin Islands have also experienced significant immigration from nearby independent states. In the Pacific approximately half the population of American Samoa has migrated from Western Samoa, and many have gone from Tonga to Niue. In some cases this represents the first stage of migration to the metropolis.

Though most island states, especially in the Pacific, have development plans, and even policies, which emphasize the need to achieve greater self-reliance, such statements are rhetoric rather than reality, a legacy of the post-independence optimism of the 1970s. In a similar manner to the situation of the DOM-TOMs, for such small island states as Kiribati, greater self-reliance is only possible 'at a price. It will not be achieved without further sacrifice in terms of foregone consumption and restricted aspirations . . . Many more sacrifices will be required in the future if a true commitment to self-reliance, is to be maintained' (Pollard 1987:23). Such sacrifices in tandem with ideological purity are rare. Despite the rhetoric of independence and self-reliance, island states, in the post-independence era, have invariably been more closely integrated into the economies of metropolitan states:

> Indeed there are diplomatic advantages to be gained by Island governments persisting with the rhetoric of autonomous development and insisting on their rights, as self-governing entities, to determine their own goals. They may thus for some time find it advantageous to refuse to recognise the MIRAB model and its implications.
> (Bertram and Watters 1985:515)

It has proved to be so; no concept has been so denigrated by Pacific island politicians and planners.

Regional co-operation and trade have largely been a failure, through complementarity and competition, hence more distant ties have increased in importance. Microstates and colonies that have a 'special relationship' with a metropolitan power are better off than those which do not. Dommen has bluntly stated that 'the particularly poor island countries are those which have failed to establish sufficiently intimate relations with a prosperous protector' (Dommen 1980:195). In the same vein Winchester has suggested that the remaining British colonies might be better off by strengthening their ties with the United Kingdom in the manner that French 'colonies' have become overseas territories and departments (Winchester 1985:309). Early in 1987 when it was feared in the Turks and

Caicos Islands that the United Kingdom might relinquish its sovereignty, there were moves there to make the colony a ward or territory of Canada (Connell 1990; 1988a:83). Because of the structure of development, the smallest states are inexorably moving towards a situation where their autonomy is severely constrained, yet none are likely to relinquish independent political status.

The states that are the greatest global per capita aid recipients are primarily colonies and territories. This has discouraged independence sentiments. Moreover, as Tuvalu and Mayotte have shown in quite different ways, secession from an independent island state substantially increases material rewards. Thorndike has even argued that the smaller Caribbean IMS, such as Grenada, 'particularly sought independence ultimately to gain access to multilateral aid funds and to participate in international forums primarily concerned with economic development, rather than from an appreciation of *its intrinsic worth*' (Thorndike 1985; my emphasis). The combination of a degree of isolation (and hence strategic significance) and a measure of political 'independence', through either sovereign status or recognition in some manner as a separate political entity, has granted superior access not only to aid but to new areas of policy formulation and concessions of other kinds. In these respects colonies have definite advantages. The reality of closer incorporation underlies most development practice and widespread high levels of migration are a reality and metaphor of development. Small is no longer beautiful; remote islands are too often 'beautiful but no place to live' (Bedford 1980:65) and by freely choosing strategies that enable the manipulation of metropolitan national policies, the structure of development of island states and dependent territories will continue to converge. As they do so the global era of decolonizaiton draws to a close. The ties that bind are likely to endure, though the conflict between freedom and equality is sure to persist.

THE SECOND MIRAGE

Over time remote islands have been more effectively incorporated into the periphery in a number of ways – including those of trade, aid, migration and political subordination.

Marriage, education and new forms of media have all stressed metropolitan ties, though rarely to the extent of Guam, where the influence of television has been so pervasive that 'not only does it make us feel homesick for places we have never been, it gives us the uneasy feeling that what we experience daily is abnormal' (Underwood 1985:171). Commodities have tied distant islands to another world; in most island states the contents of the stores are imported; in rural Fiji of 82 products sold at one small island co-operative in mid-1986 a third were food products and well over half the goods originated entirely outside Fiji, 'an astonishing testimony to the history of colonialism and the more recent organization of international commerce', the result of becoming insatiable consumers, conditioned to need an ever-increasing array of disposable goods' (Price 1985:217; McInnes and Connell 1988). The store, an incursion from another world, has incorporated remote islands into that other world even more effectively than production or migration. Incorporation has emphasized relative deprivation, and the manner in which island states can never have the range of social and economic opportunities of metropolitan states.

The exotic image of tropical islands and widespread assumptions of something akin to history of 'subsistence affluence' (Fisk 1982), even enabling some to discover situations 'where the poor are happy' (Owen 1955), contrasts with a reality of struggle for survival, and the erosion of sovereignty, as self-reliance becomes no more than a chimera. Even in much larger countries attempts to achieve self-reliance often appear no more than reflections of the aspirations that must suffice if growth cannot easily be achieved: as Joseph has put it, in a Nigerian context, self-reliance is 'little more than a ritual for exorcising the devil of dependence' (Joseph 1978:223). This is not very different from Leys's observation for Kenya that foreign capital must 'be *ritually* "humiliated" while practically wooed' (Leys 1985:208). Yet because of the difficulty of becoming self-reliant in such basic requirements as food and energy, even if appropriate policies were chosen, put most bluntly,

it would only be by accepting primitive standards of development for all the people that autarky could be made in any way practicable . . . Even in the Pacific, perhaps the one remaining part of the world in which such ideas do not appear absurd, the pace of absorption into the world economic and political system is quickening all the time.

(Payne 1987:56)

But then islands and societies were never wholly isolated and self-reliant in the past. Where this tended to occur, islands like Anuta, despite complex precautions, were rare examples of the 'Malthusian crisis' and historic populations on islands such as Pitcairn perished. Consequently, as in the Caroline Islands of Micronesia, atolls organized themselves into coral clusters and complexes, for distant reciprocal co-operation. It could not always ward off devastation. There is even less reason to believe that island states can now achieve greater self-reliance. The future lies in maximizing the various elements of interdependence, to take advantage of concessionary schemes and the strategic significance of all IMS whilst strengthening self-reliance in such critical areas as food production and artisanal fisheries, in order to maintain the 'subsistence safety-net'.

Quite recently a review of development problems in island states concluded by posing the question: 'Is it too much to suggest that small islands, for all the problems and constraints that confront them, could become the laboratory in which alternative development strategies, shaped by the notion of self-reliance, first see the light of day?' (Dolman 1985:63). Unfortunately for populist notions, such a romantic vision has arrived too late. Indeed it points directly to a second mirage – that there really is something that can be identified as development. There is not. Development is a wholly relative concept. Thus whilst Wallman's early conceptualization of development emphasized both basic needs and autonomy, she stressed that they were not complementary processes. In IMS so thoroughly incorporated into global economy and society, even improvements in basic needs are set against their superior satisfaction elsewhere; development may have occurred but it has not been achieved. At the same time autonomy has certainly declined. In the three senses that Galtung has defined power – 'ideological power is the power of ideas. Remunerative power is the power of having goods to offer, a 'quid' in return for a 'quo'. Punitive power is the power of having 'bads' to offer, destruction; also called force, violence' (Galtung 1973:33) – the IMS have experienced a decline. In Seers's terms, they have not gained in independence, despite the 'culture of resistance' that has, in some contexts, enabled the retention of an 'independence of spirit' (Petersen 1984:359).

As one ideological world becomes dominant, and even nationalism is constructed in European terms, then the retreat to self-reliance will become less and less palatable. Planners brought their prescriptions for development to the IMS as they had previously done in other parts of the world. Over twenty years ago van Arkadie wrote of these 'gifts of the latest advice' that 'we might well see these fluctuations as more geared to meeting the needs of the now considerable development studies profession than responding to real problems – a sort of planned obsolescence in ideas' van Arkadie (1978:409). Such heretical views were expressed in a different form by Fisk:

> We have all seen the rise and fall of fashions in cures for underdevelopment. At one time it was the transfer of technology; at another the provision of capital; at another it was the development of cooperative societies; then it was economic planning; then there was a fashion for community development; more recently there has been a fashion for what is called integrated development.
>
> (Fisk 1978:371)

Such fashionable changes have continued. Underlying these sometimes conflicting prescriptions were the visions of Europeans – who sought to impose variants of tradition and authenticity on latter-day 'noble savages'. Yet as such notions challenged island conceptions of the good life, and the 'passing parade of paradigms' (Baker 1979:167) turned to irrelevance, IMS have sought to construct a world of choices, a negotiated multivariate dependence.

By freely choosing development strategies which enable the manipulation of metropolitan national policies, IMS will continue to 'live with some degree of uncertainty, but with insurance provided by the realities of geopolitics' (Bertram 1986:821). Diversity is the key to development, but a diversity that is global and no longer local. For all those who have seen in small islands 'laboratories of development', many more have predicted the demise of island populations. Recently Ward concluded a keynote paper on the Pacific as follows:

> Perhaps a hundred years hence . . . almost all the descendants of today's Polynesian or Micronesian Islanders will live in Auckland, Sydney, San Francisco and Salt Lake City. Occasionally they may recall that their ancestors once lived on tiny Pacific Islands. Even more occasionally they may visit the resorts which, catering for scuba divers, academic researchers or gamblers may provide the only permanent human activities on lonely Pacific Islands, set in an empty ocean.
>
> (Ward 1989:245)

It is a scarcely alluring vision but even a precipitate greenhouse effect will leave it no more than that. The astonishing manner in which Pitcairn islanders have clung to their tiny isolated outpost of empire (Connell 1988b) is symbol and substance of the new island states in which identity, combined with isolation and strategic location, have shaped a new world where multifaceted dependence might be transformed into aid with dignity. But nowhere can it be transformed into development.

NOTE

A longer version of this chapter appears in *The Contemporary Pacific* 3 (1991).

REFERENCES

AIDAB (1987) *Australia's Relations with the South Pacific*, Canberra: Australian Government Printing Service.

Aldrich, R. and Connell, J. (1991) *France's Overseas Frontier*, Melbourne: Cambridge University Press.

Baker, R. (1979) 'Review of *Why Poor People Stay Poor*', *Journal of Modern African Studies* 18: 167-73.

Barry, T., Wood, B. and Preusch, D. (1984) *The Other Side of Paradise*, New York: Grove Press.

Beckett, J. (1987) *Torres Strait Islanders. Custom and Colonialism*, Cambridge: Cambridge University Press.

Bedford, R.D. (1980) 'Demographic processes in small islands: the case of internal migration', in H.C. Brookfield (ed.) *Population-Environment Relations in Tropical Islands: The Case of Eastern Fiji*, Paris: UNESCO, 27-59.

Bertram, G. (1986) ' "Sustainable development" in Pacific micro-economies', *World Development* 14: 809-22.

Bertram, G. and Watters, R. (1985) 'The MIRAB economy in South Pacific microstates', *Pacific Viewpoint* 26: 497-519.

Britton, S. (1987) 'Tourism in Pacific Island states: constraints and opportunities', in S. Britton and W.C. Clarke (eds) *Ambiguous Alternative, Tourism in Small Developing Countries*, Suva: University of the South Pacific, 113-39.

Brookfield, H.C. (1975) *Interdependent Development*, London: Methuen.

Brookfield, H.C. (1987) 'Export or perish: commercial agriculture in Fiji', in M. Taylor (ed.) *Fiji: Future Imperfect*, Sydney: Allen & Unwin, 46-57.

Charle, E.V. (1986) 'Foreign trade patterns and economic development in the South Pacific', *Journal of Pacific Studies* 12: 1-32.

Chenery, H., Ahluwalia, M.S., Bell, C.L.G., Duloy, J.H. and Jolly, R. (1974) *Redistribution with Growth*, Washington: World Bank..

Conklin, E.C. (1987) 'Caribbean Basin Initiative. A regional solution for America's threatened enterprise?' *Focus* 37: 2-9.

Connell, J. (1980) 'Rural development: green, white, red or blue revolutions?' in J. Friedmann, T. Wheelwright and J. Connell (eds) *Development Strategies in the Eighties*, Sydney: Development Studies Colloquium Monograph no. 1.

Connell, J. (1987) 'Bermuda: a failure of decolonisation?', University of Leeds, School of Geography, Working Paper no. 492.

Connell, J. (1988a) *Sovereignty and Survival: Island Microstates in the Third World*, University of Sydney, Research Mongraph no. 3.

Connell, J. (1988b) 'The end ever nigh: contemporary population change on Pitcairn island', *Geo Journal* 16: 193-200.

Connell, J. (1988c) 'Contemporary issues in island micro-states', in M. Pacione (ed.) *The Geography of the Third World: Progress and Prospect*, London: Routledge, 427-62.

Connell, J. (1990) 'The Turks and Caicos islands: beyond the quest for independence', *Caribbean Geography* 3(1): 53-62.

Connell, J. (1991) 'The new Micronesia: pitfalls and problems of dependent development', *Pacific Studies* 14(2): 87-120.

Dolman, A.J. (1985) 'Paradise Lost? The past performance and future prospects of small island developing countries', in E. Dommen and P. Hein (eds) *States, Microstates and Islands*, London: Croom Helm, 40-69.

Dommen, E. (1980) 'External trade problems of small islands states in the Pacific and Indian Oceans', in R.T. Shand (ed.) *The Island States of the Pacific and Indian Oceans*, Canberra: Development Studies Centre Monograph no. 23, 179-99.

Dommen, E. and Hein, P. (1985) 'Foreign trade in goods and services: the dominant activity of small island economies', in E. Dommen and P. Hein (eds) *States, Microstates and Islands*, London: Croom Helm, 152-84.

Fairbairn, I. (1985) *Island Economies*, Suva: Institute of Pacific Studies.

Fisk, E.K. (1978) 'Traditional agriculture and urbanization: policy and practice', in E.K. Fisk (ed.) *The Adaptation of Traditional Agriculture*, Canberra: Development Studies Centre Monograph No. 11, 345–77.

Fisk, E.K. (1982) 'Subsistence affluences and development policy, *Regional Development Dialogue* (Special Issue) 1–12.

Galtung, J. (1973) *The European Economic Community: An Emerging Superpower*, London: Allen & Unwin.

Gauhar, A. (198) 'Manley rides the new wave', *South* 105: 10–11.

Hart, K. (1974) 'A model of development to avoid', *Yagl-Ambu* 1: 8–15.

Hein, C. (1986) 'Export processing zones in island countries: the case of Mauritius', unpublished paper, Martinique.

Hezel, F.X. and Levin, M.J. (1990) 'Micronesian emigration. Beyond the brain drain', in J. Connell (ed.) *Migration and Development in the South Pacific*, Canberra: Pacific Research Monograph no. 24, 42–60.

Jackson Report (1984) *Report of the Committee to Review the Australian Overseas Aid Program*, Canberra: Australian Government Publishing Service.

Jones, H. (1989) 'Mauritius: the latest pyjama republic', *Geography* 74: 268–9.

Joseph, R.A. (1978) 'Affluence and underdevelopment: the Nigerian experience', *Journal of Modern African Studies* 16: 221–39.

Leys, C. (1985) *Underdevelopment in Kenya*, London: Heinemann.

Lea, D. (1980) 'Tourism in Papua New Guinea: the last resort', in J.M. Jennings and G.J.R. Linge (eds) *Of Time and Place*, Canberra: Australian National University Press, 211–31.

Lesourd, M. and Reaud-Thomas, G. (1987) 'Le fait créole dans la formation de l'identité nationale en Republicque du Cap-Vert', in J.P. Doumenge *et al.* (eds) *Isles Tropicales, Insularite, Insularisme*, Talence: Centre de Researches sur les Espaces Tropicaux, 107–24.

Lipton, M. (1977) *Why Poor People Stay Poor*, London: Temple Smith.

McDonald, I. (1989) 'The destiny of a small nation', *Caribbean Affairs* 2(3): 173–9.

Macpherson, C. (1990) 'Stolen dreams. Some consequences of dependency for Western Samoan youth', in J. Connell (ed.), *Migration and Development in the South Pacific*, Canberra: Pacific Research Monograph no. 24, 107–19.

McInnes, L. and Connell, J. (1988) 'The world system in a Fijian store', *South Pacific Forum* 4: 116–21.

Miles, W.F.S. (1986) *Elections and Ethnicity in French Martinique. A Paradox in Paradise*, New York: Praeger.

Momsen, J. (1986) 'Linkages between tourism and agriculture:

problems for the smaller Caribbean economies, University of Newcastle, Department of Geography Seminar Paper no. 45.

Munro, D. (1990) 'Migration and the shift to dependence in Tuvalu', in J. Connell (ed.) *Migration and Development in the South Pacific*, Canberra: Pacific Research Monograph no. 24, 29–41.

Naidu, V. (1981) 'Fijian development and national unity', *Review* 2(5): 3–12.

Owen, R. (1995) *Where the Poor are Happy*, London: Travel Book Club.

Paul, E. (1987) *Fisheries Development and the Food Needs of Mauritius*, Rotterdam: Balkema.

Payne, T. (1987) 'Economic issues', in C. Clarke and T. Payne (eds) *Politics, Security and Development in Small States*, London: Allen & Unwin, 50–61.

Petersen, G. (1984) 'The Ponapean culture of resistance', *Radical History Review* 28–30: 347–66.

Pollard, S. (1987) 'The viability and vulnerability of a small island state: the case of Kiribati, Islands?'. Australia Working Paper no. 87/14, National Centre for Development Studies, Canberra.

Price, R. (1985) 'The dark complete world of a Caribbean store', *Review* 9: 215–18.

Ramassamy, A. (1987) *La Réunion, decolonisation et integration*, St. Denis: AGM.

Rogers, R.F. (1988) *Guam's Commonwealth Effort 1987–8*, Guam: Micronesian Area Research Center.

Rubenstein, H. (1988) 'Ganja as a peasant resource in St. Vincent', in J.S. Brierley and H. Rubenstein (eds) *Small Farming and Peasant Resources in the Caribbean*, Winnipeg: Manitoba Geographical Studies no. 10, 119–33.

Sathiendrakumar, R. and Tisdell, C. (1986) 'Fishery resources and policies in the Maldives', *Marine Policy* 10: 279–93.

Schutz, B. and Tenten, R. (1979) 'Adjustment, problems of growth and change', in A. Talu (ed.) *Kiribati. Aspects of History*, Suva: Institute of Pacific Studies, 106–27.

Seers, D. (1969) 'The meaning of development', *International Development Review* 11: 2–6.

Seers, D. (1977) 'The new meaning of development', *International Development Review* 19: 2–7.

Shankman, P. (1976) *Migration and Underdevelopment. The Case of Western Samoa*, Boulder: Westview Press.

Taylor, M. (1987) 'Issues in Fiji's development: economic rationality or aid with dignity? in M. Taylor (ed.) *Fiji: Future Imperfect*, London: Allen & Unwin, 1–13.

Thorndike, T. (1985) *Grenada, Politics, Economics and Society*, London: Frances Pinter.

United Nations Conference on Trade and Development (UNCTAD) (1985) *The Least Developed Countries, 1985 Report*, London: United Nations.

Underwood, R.A. (1985) 'Excursions into inauthenticity: the

Chamorros in Guam', in M. Chapman (ed.) *Mobility and Identity in the Island Pacific, Pacific Viewpoint* 26(1): 160–84.

Van Arkadie, B. (1978) Town versus country? *Development and Change* 8: 409–15.

Wallman, S. (1977) *Perceptions of Development*, Cambridge: Cambridge University Press.

Walters, H.D. (1984) 'Fisheries development in St Lucia and the Lesser Antilles', *The Courier* 85: 94–5.

Ward, R.G. (1989) 'Earth's empty quarter? The Pacific islands in a Pacific century', *Geographical Journal* 155: 235–46.

Wendt, A. (1987) 'Western Samoa 25 years after: celebrating what?' *Pacific Islands Monthly* 58: 14–15.

Winchester, S. (1985) *Outposts*, London: Hodder & Stoughton.

World Bank (1987) *Fiji. A Transition to Manufacturing*, Washington: World Bank.

World Bank (1990) *World Development Report 1990*, Washington: World Bank.

Part II

EMPIRICAL INVESTIGATION OF DEVELOPMENT ISSUES

8

THE LEGACY OF COLONIALISM

The experience of Malta and Cyprus

Henry Frendo

INTRODUCTION

This chapter will look briefly at Malta and Cyprus, two strategically-located Mediterranean islands which also share a British colonial past, indicating some similarities and differences in their development. As independent states, too, in spite of some marked contrasts, these islands continue to have certain common concerns and aspirations in the broader context of a European Mediterranean.

Much as interdisciplinary approaches to historiography exemplified by Fernand Braudel have been influential in regarding the Mediterranean as a regional entity (Braudel 1949), this sea – largely due perhaps to its very geography – has been perceived in 'holistic' terms since time immemorial. Greeks from Homer to Herodotus characterize it simply as 'The Sea'; in the Old Testament it is the 'Great Sea' (*Mare Magnum*); and Romans later called it 'Our Sea' (*Mare Nostrum*), as did others in more recent times. The epithet *Mare Mediterraneum* literally means 'the sea in the middle of the earth'. Samuel Johnson even said that 'almost all that which sets us above savages has come to us from the shores of this sea' (Vella 1985). Toynbee, however, categorized Europeans as

> those inhabitants of the north-western peninsula of the Old World, and of the adjacent islands, who are ecclesiastical subjects or ex-subjects of the Patriarchate of Rome: in other words we mean those Catholic and Protestant

151

Christians who live in the north-western corner of the Old World.

(Beloff 1957)

As a human unit the Mediterranean provided a rather closed area for exchange and intercourse, but it has also been 'the great divider, the obstacle that had to be overcome'. It is cities and communications, writes Braudel, that have imposed 'a unified construction on a geographical space' (Braudel 1973).

The history no less than the geography are in the sinews of this complex area, whether we look to the metropolitan hinterland cities of Europe or to the Mediterranean basin's shores, or, indeed, if we consider both at the same time – thus debunking' conventional notions of defined continents and creating a new inter- or extra-continental focus (Frendo 1981).

The Mediterranean is a 'Middle Sea' in a double role: not only does it lie between Europe and Africa, it also has joined the Atlantic to the Near and the Far East when through the Straits of Gibraltar and, after 1869, through the Suez Canal, it became the main trading route for the imperial powers who, until fifty years ago, largely controlled its destinies: the British, the French and, to a lesser extent, the Italians.

Throughout this period Mediterranean islands were sought after and changed hands: Minorca and the Ionian Islands, Crete and Rhodes, Corsica and Sardinia. At the crossroads of traffic in all directions were three main centre points: Gibraltar, Malta and Cyprus – all three coloured red on the map in worn school textbooks. After holding on to Gibraltar in repeated sieges, having captured it from Spain in 1704, the British extended their hold eastwards, taking Malta from France in 1800, then Cyprus from Turkey in 1878. The first remains so far a British possession; the second only became independent in 1964 but survives as a unitary democratic state; the third retains British military bases and its 1960 independence arrangement has been rendered unworkable.

Insularity and relative isolation counterbalanced by strategic geographical location, as demonstrated by long stints of foreign control, endowed such Mediterranean contexts and communities with remarkable identities and peculiarities, despite their own serious reservations about the respective metropoles – Madrid or Paris, Rome or London, Athens or Ankara. In the

Gibraltar-Malta-Cyprus axis, through the wand of imperialism Britain could be said to have superimposed common unities: teaching the (European) natives English, promoting commercial and military activity thereby reorienting demand for labour and skill, and introducing (at least *pro forma*) tenets of liberal constitutional government and procedural legal maxims. At a deeper level, British colonialism preyed on existing or potential differences or conflicts of interest or of perception, frequently inducing an Englishness that was unreal, a rather transient hybrid in place of what might have been a less contorted evolution.

In both areas under review, the overseas Anglo-Saxon and the generally Latin standard came into contact, were contrasted, sometimes adapting in collaboration, more often clashing through resistance and opposition. Rulers tried to create or draw upon a 'non-European' or 'less-Latin' residue as a malleable counterweight to the usually dominant anti-English 'traditionalist' formation. At the same time, and partly for such reasons, opponents stretched their hands northwards towards real or imaginary motherlands for cultural – and indeed, at times, political – sustenance and support. Scars ran deep: the result was a (still unresolved) self-identity crisis in the 'British' Mediterranean.

Malta and Cyprus in varying degrees have long sought to prove their Europeanity. As independent states they have been increasingly on the same wavelength as the state visit to Malta of Cypriot President George Vassiliou in June 1991, and an earlier visit by Archbishop Makarios, have further confirmed. A cultural agreement signed in February 1991 is similarly indicative of growing contacts between these two island states and ex-colonies.

MALTA

Malta was under British rule for 164 years. During this period, especially from the 1870s onwards, the British had sought to anglicize and, so far as was feasible, to assimilate the Maltese. This was mainly out of fear that the cultural, religious, literary, historical and geographical affinities and proximities with the Italian peninsula, threatened imperial interest potentially or actually. 'Italianita' thus became a *cheval de bataille* of the emergent colonial nationalism in resistance to assimilation and acculturation: 'denationalization' was a favourite term used. Arguing that the British Empire would end but the geography would not change, 'pro-Italians' sought to restate and reinvigorate precisely those links and sensitivities which their rulers wished to dampen and to fade out, thereby reinforcing them for a long while in that process. While the socially-powerful Roman Catholic church hierarchy were treated with silken gloves, bishops successively knighted and Protestant proselytization kept to a minimum, what the British also did was to instigate and to patronize the growing 'pro-English' party, holding out preferment, or simply opportunity, in jobs at the dockyard, with the forces and in the civil service (provided those employed could speak some English and preferably were known to be 'loyal'). Emigration prospects to the English-speaking world, especially Australia after the Second World War, were also greatly encouraged. In effect this helped to alienate the working classes, who tended to be bread-and-butter loyalist, from the middle classes who, being better educated and more independent of means, tended to oppose colonialism and to seek responsible government for the island (Frendo 1979; 1989).

In the ensuing occupational and educational, parochial and national, governmental and partisan confusion, the nationalist party essentially upheld Italian against its substitution by English, and looked to Italy for cultural survivance rather than to Britain, whereas the imperialist and later the labour parties tended to look to Britain and continuing dependence on the Crown for utilitarian, socio-cultural and emigration links. The local camp was thus divided into a generally internecine rivalry.

Not so small as Gibraltar, nor ethnically and religiously divided like Cyprus, Malta had a fertile enough ground for the evolution of a common nationality and the development of democratic structures (Frendo 1975). That as late as the 1950s its Labour Party, led by Dominic Mintoff, should see as a long-term solution Malta's integration with the United Kingdom possibly betrays a rootless want of national belonging, or, more simply, the poor countryman's bid for a quick penny, or both together (Austin 1971). An anglophile in the early part of his career, Mintoff, like Lord Strickland before him, was in this sense the obverse of what the elder and the younger Mizzi had been in the nationalist movement: *in extremis* one had a progression from apparently being 'more Italian than the Italians' to 'more English than the English'. Not unlike Enrico (Nerik) Mizzi who, in 1912, had proposed federation with Italy, Mintoff saw the need for dependence and association under certain conditions, but he expressed this differently – desiring to deepen rather than to dampen the British connection. In spite of all the latter-day anti-British rhetoric, when Mintoff returned to power in 1971 he had the Anglo-Maltese defence agreement renegotiated and extended to 1979. In anticipation of the departure of his onetime mentors, however, he came to rely increasingly on other outside support – especially Gaddafi's Libya.

While looking beyond to the geographically-more-removed Anglo-Saxon standard, however, the Labour Party had supported the elevation of Maltese, a Semitic vernacular, to the position of a national language and as a tool for the learning of English. Learning English through the medium of Maltese was a wedge that eventually drove Italian from the curriculum and, after 1940, from public affairs (until the advent of Italian television in the late 1950s). Maltese and English remain the official languages today. Yet, once the 'integration with Britain' failed, Mintoff's party in 1958 proposed a 'Break with Britain' resolution, which Borg Olivier's Nationalists, who had opposed integration, readily supported. Later on, in the 1970s and 1980s, from government, the Labour Parties upheld 'neutrality' and 'non-alignment' and shied away from conventional western associations. The Nationalists, on the other hand stuck to a pro-European policy and in 1976 first committed the party to seeking full membership of the EEC. In July 1990, once back in government, they applied for membership, as Cyprus also did.

The Labour or Socialist Party was in office from 1971 to 1987 far the longest administration in Malta's electoral history. These were difficult and trying times which have been commented on extensively in the world media. But in May 1987 power was transferred to the Nationalist (Christian Democratic) Party led by Edward Fenech Adami, who won an absolute majority of the popular vote for the second time in succession.

CYPRUS

The Cypriot experience as an island-state and as a 'nation' has been bitter and has not so far seen any lasting redress, even as efforts to bring about a solution intensify with prodding from President Bush in both Athens and Ankara.

In January 1985, at the UN Headquarters in New York, the then Cyprus President, and former Makarios confidante, Kyprianou, met the Turkish Cypriot leader Denktash for the first time since 1979. This 'summit' was intended to produce a framework for a comprehensive settlement of the post-1974 Cyprus question: as a timetable for implementing the phases leading to a federated republic of Cyprus. The parties, however, could hardly agree about an agenda and the terms of reference to be adopted.

Ironically, long-time rivals Clerides and Denktash were both British-trained lawyers – one from Lincoln's Inn, the other from Gray's Inn. But differences in Cyprus run deeper than in either Malta or Gibraltar. As in Lebanon, internal cultural conflicts were exacerbated by ethnic as well as by religious differences.

The one unifying colonial factor in a positive sense, in this situation, may be said to have been the common usage of English between the two communities, but the British never set up a university in Cyprus during their eighty-two years there. It is only now, in the early 1990s, that one is being launched – with Greek and Turkish as its main languages.

In the wake of an attempted *coup d'état*, inspired by mainland Greek officers against President Makarios in 1974, which sought crudely to impose the long-held Greek Cypriot aspiration of *enosis* or union with Greece, Turkey invaded the northern part of Cyprus, invoking the 1960 tripartite Treaty of Guarantee, and refusing to budge ever since. These two events finally split the island in half, with Turkey dominant in the north, and the Greek Cypriots concentrated more than ever in the central and southern section. Feelings have run so high that it is not clear whether outright partition can be averted, or whether a semblance of unity in this island state may be saved through federation. The Turkish side is isolated, with the Turkish-Cypriot state not recognized by any other state save Turkey herself, and the Turks have been desperately trying to justify their position (Denktash

1982), or at least to explain it historically and juridically (Necatigil 1989).

United Nations (UN) Security Council Resolution 649 of 12 March 1990 again called for 'a bi-communal Federal Republic of Cyprus' safeguarding its independence and territorial integrity and excluding 'in whole or in part with any other country and any form of partition or secession'. The leaders of the two communities were urged to seek 'a federation that will be bi-communal as regards the constitutional aspects and bi-zonal as regards the territorial aspects' (Ertekun 1990).

Britain first assumed control of Cyprus by the 1878 Treaty of Berlin, then annexed it in 1914, and then made it a 'Crown Colony' in 1925. But from the start the predominant feeling in the island was 'pro-Greek'; indeed, three-quarters of the population were of Greek descent and spoke Greek. Ironically, however, the island was geographically tucked away under Turkey's belly, and politically had most recently belonged to it, not Greece. Nevertheless, the religious, ethnic and linguistic ties with Greece were dominant, as was fear or resentment of the former Ottoman power.

The Greek-Cypriot bishop of Citium told Sir Garnet Wolseley in 1878 that the Cypriot people welcomed the occupation because they expected Britain would let them become part of 'Mother Greece', as had happened with the Ionian Islands some years earlier (and indeed, this might well have happened in 1915 during the First World War when Britain sought to appease Greece and defend Serbia against Bulgaria). Nevertheless, Greeks and Turks in Cyprus for the most part co-existed: it was not uncommon for minorities from either side to live peacefully in mixed villages all over the island (Salih 1978).

In reaction to the anti-British movement for union with Greece, or for independence from Britain, the British tended naturally to cultivate the greater fidelity of the Turkish-Cypriots who, being in the minority, stood more to gain from British patronage. Seeds of additional division were thus sown under the noble pretext of ensuring minority rights, which truly may have sometimes required protection against the EOKA organization of General Grivas (Reddaway 1986; Hitchens 1984). There was less protection by mutual consensus; colonial despatches written in London for application in the field did not always help the intended beneficiaries in the long run.

While not perhaps as disquieting as Lebanon, and not as continuously bloody, the Cypriot question is none the less a tragic one because with goodwill, without undue interference, and within a homegrown constitutional framework, it is known that different ethnic groups with less of a shared past than Cypriots have co-existed and developed a common adherence and allegiance to statehood, even to nationhood. While not denying the persistence of a strong separate consciousness – of being part of the 'Greek' and 'Turkish' nation respectively – students of Cypriot history have sometimes found astonishing 'not the existence but the weakness of mutual distrust' (Kadritzke and Wagner 1979).

Yet in a way Cyprus never really became fully independent. Independence was itself subject to the tripartite agreement reached in 1959 between Britain, Greece and Turkey, with two sovereign military bases still retained by Britain. Greece and Turkey, although both members of NATO, continued keenly eyeing every turn of events in their respective spheres of influence and, in the end, claiming the right to intervene under the very agreement which was supposed to guarantee the country's independence. Cypriots were repeatedly excluded from negotiations by the 'powers' on Cyprus (Erlich 1974). Even if a separate national consciousness incorporating the Greek and Turkish strains had been permitted to develop, which it was not, Cyprus would still have been subjected to political interests emanating from its strong and rival neighbours. Both felt they had a stake in it, if only through the provenance of their people(s) inhabitating the place.

CONCLUSION

Thus, on the one hand we have the *unities* of small Mediterranean states – common trade routes and cultural import-export, ethnic admixtures over the centuries, the ancient heritage, civilizations and treasures, the subjugation to 'foreign' rule and, increasingly, the general resistance to that with a view to independence, prosperity and peace. On the other hand – is this insidiously an ultimate 'unity'? – you have the very considerable problems facing these southern Europeans, at one time prototypical and peripheral, in securing their independence politically, economically and, not least of all, culturally. Diversities seem to reinforce the unities: would incorporation and integration in the dreamed-of larger whole – be it tomorrow's Europe or yesterday's Empire – secure and refine these nationalities or swamp and gradually eradicate them?

As European union should have neither the imperialistic motive nor the nationalistic itch, it could well serve to redirect latent feelings of dispossession, insecurity, and unfulfilment experienced in areas such as Cyprus and Malta, by affording a broader and more accommodating body politic willing to shape a new economic order and yet inspired by a common heritage. A European Mediterranean provincialism might be better able to guard and to cherish the individual charms composing it.

In the history of the British empire Malta and Cyprus were the only two island colonies with strong historical and cultural claims to Europeanity. They were in effect the only two European British colonies over a long period, although in the case of Cyprus there was always the Turkish angle. Although somewhat distanced by British occupation from the continental European mainstream, and geographically 'emarginated' on the southern fringes of the continent, especially Cyprus, generally both felt they were spiritually and culturally part of mainland European 'mothers'; that British colonialism defied or denied what to some of their politicians, mostly in Cyprus, appeared like manifest destiny. Once they had obtained independence in the early 1960s and tried to put their house in order, both countries have sought to vindicate those past self-fulfilling claims by joining the European Community as full members.

REFERENCES

Austin, D. (1971) *Malta and the End of Empire*, London: Cass.

Beloff, M. (1957) *Europe and the Europeans*, London: Council of Europe, 7.

Braudel, F. (1973) *The Mediterranean and the Mediterranean World in the time of Philip II*, Volume 2 London: Collins, 276–7.

Denktash, R.R. (1982) *The Cyprus Triangle*, London: Allen & Unwin.

Erlich, T. (1974) *Cyprus 1958–67*, Oxford: Oxford University Press.

Ertekun, N.M. (ed.) (1990) *The Status of the Two Peoples in Cyprus*, Lefkosa: Public Information Office, Turkish Republic of Northern Cyprus.

Frendo, H. (1975) 'Language and nationality in an island colony: Malta', *Canadian Review of Studies in Nationalism* 3(1): 22–33.

Frendo, H. (1979) *Party Politics in a Fortress Colony. The Maltese Experience*, Valletta: Midsea.

Frendo, H. (1981) 'Mediterranean irredentism', *Perspektiv* 4(15): 28.

Frendo, H. (1989) *Malta's Quest for Independence. Reflections on the Course of Maltese History*, Valletta: Dougall.

Garratt, G.T. (1939) *Gibraltar and the Mediterranean*, London: Jonathan Cape, 43.

Hitchens, C. (1984) *Cyprus*, London: Quartet.

Katritzke, N. and Wagner, W. (1979) 'Limitations of independence in the case of Cyprus', in P. Worsley and P. Kitromilides (eds) *Small States in the Modern World: The Conditions of Survival*, Nicosia: The New Cyprus Association, 112.

Necatigil, Z.M. (1989) *The Cyprus Question and the Turkish Position in International Law*, Oxford: Oxford University Press.

Reddaway, J. (1986) *Burdened with Cyprus: The British Connection*, London: Weidenfeld & Nicolson.

Salih, H.I. (1978) *Cyprus: The Impact of Diverse Nationalism on a State*, Alabama: University of Alabama Press.

Vella, A.P. (1985) 'Mediterranean Malta', *Hyphen* 4(5): 169.

9

THE SMALL ISLAND FACTOR IN MODERN FERTILITY DECLINE

Findings from Mauritius

Huw Jones

Is there such a thing as small island demography? Within the modern Third World context, it has sometimes been suggested that small islands have relatively low fertility, low mortality and high rates of migration to developed countries. This does seem to accord with recent demographic experience (Barbados, Jamaica, Trinidad, Mauritius, Fiji, Singapore spring readily to mind), and in the case of fertility, at least, this has been empirically confirmed (Mauldin and Berelson 1978).

There are two likely causes of such demographic distinctiveness: first, the circumscribed space of small islands provides logistical advantages both for the eradication of disease and the provision of health care and family planning facilities. Moreover, a combination of high population density and restricted space encourages the population to consider not only emigration but also voluntary constraints on family size. Tikopia in the Solomon Islands provided a classic illustration of cultural mechanisms adopted by a traditional society to curb population growth by combinations of withdrawal, abortion, delayed marriage, celibacy and infanticide (Firth 1957). In such circumstances there is a recognizably *collective* interest in fertility restraint, and modern governments can harness this in the implementation of family planning programmes. This contrasts with larger Third World countries where there is often a distinction between the macro- and micro-consequences of high fertility. There, at a society level, high fertility invariably entails high net costs, but at a family level net costs are likely to be low, and there may well be net benefits.

The second cause relates to the fact that mortality and fertility reduction are promoted by the relatively high development levels that characterize most small island states. Such development levels are, in part, *caused* by island status, stemming from the period during which many small islands were part of the European maritime system when the West did not have the capability to penetrate and transform continental areas (Caldwell *et al*. 1980). Another contributory factor must be the higher levels of development aid per capita enjoyed by microstates (the great majority of which are islands) for whatever reason (historic links with the west, 'purchase' of UN votes, and so on). Even after standardizing for development level, a clear relationship between small island status and demographic modernization persists within the Third World (Mauldin and Berelson 1978).

The key causal relationships may be summarized below in Figure 9.1.

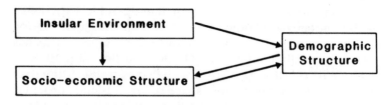

Figure 9.1 The relationship between small island status and demographic modernization.

PLANTATION-ISLAND DEMOGRAPHY

Former plantation islands, sometimes still characterized by large-scale, monocultural, crop production for export, may well form a demographic subtype based on the following common characteristics:

(a) massive labour immigration during the creation of the plantation economy;
(b) ethnic diversity, stemming from the different geographical origins of immigrant groups;
(c) a rural proletarian society (Lamusse 1980);

162

(d) initial encouragement of high fertility for labour needs;
(e) volatility in marriage rates and other demographic indicators in relation to boom/slump export-crop cycles;
(f) rural population nucleation in estate villages;
(g) the impact of westernization directly on modern mortality reduction and indirectly on fertility reduction.

Support for this hypothesis of a plantation-island demography is provided by Cleland and Singh's (1980) analyses of two regional data sets. They find a clear positive correlation between small island status and modern fertility decline among Latin American-Caribbean countries, but *not* among Pacific-East Asian countries; it is, of course, in the former region that plantation islands are overwhelmingly concentrated.

It could well be, therefore, that it is the coincidence of small island status and a plantation-type economy that particularly promotes demographic modernization. Such coincidence is exemplified by Mauritius – a very small, mid-Oceanic island, whose economy until very recently has been almost totally reliant on the cultivation, processing and export of sugar.

MAURITIUS: RECENT DEMOGRAPHIC EVOLUTION

In the early 1960s the Indian Ocean island of Mauritius was widely regarded as suffering from one of the world's gravest problems of population pressure on limited resources. The post-war eradication of malaria was the basis of a rapidly-falling death-rate (Sombo and Tobutin 1985) and an increasing birth-rate (Figure 9.2), so that by the 1962 Census the population of 682,000 had reached a density of 370 per sq. km. The annual growth-rate in population of just over 3 per cent was posing acute problems in an island whose economy was almost totally reliant on sugar, where the natural limits to cane cultivation had already been reached and where cyclone damage to crops and infrastructure was often devastating. By the mid-1980s the population growth-rate had been reduced to almost 1 per cent per year and fertility rates to below replacement level (Table 9.1), making Mauritius one of the most outstanding cases of modern fertility reduction in the Third World.

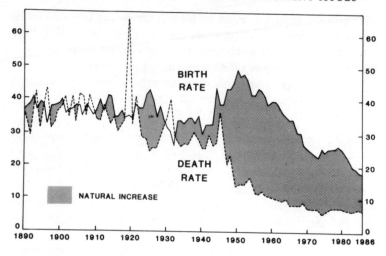

Figure 9.2 Birth-rates and death-rates (per 1,000), Mauritius, 1890–1986

Table 9.1 Mauritius: selected demographic characteristics, 1962–86

Year	Census population	Crude birth-rate*	Crude death-rate*	Rate of natural increase*	Infant mortality rate*	Total fertility rate, per woman
1962	681,619	38.5	9.3	29.2	60.1	5.86
1972	826,199	24.8	7.9	16.9	63.9	3.42
1979		27.2	7.2	20.0	32.9	3.07
1983	968,609	20.6	6.5	14.1	25.6	2.20
1986		18.3	6.7	11.7	26.3	1.94

* per 1,000

Note: The state of Mauritius also includes Rodrigues, an island of some 34,000 population 560km east of the island of Mauritius. All the data and discussion in this paper refer only to the island of Mauritius.

Most macro-level, cross-sectional studies of fertility among less-developed countries conclude that variation is closely related to a combination of development, cultural and programme variables (Mauldin and Berelson 1978; Tsui and Bogue 1978; Freedman and Berelson 1976). In other words, countries that have achieved substantial fertility declines owe these to some combination of high development level, favourable cultural background and strong family-planning programme effort. It is useful, therefore, initially to consider how well Mauritius conforms to this global model by evaluating its potential for fertility decline on each of the three dimensions.

Development level

Table 9.2 provides some objective data on several conventional indicators of development or modernization around 1970, an appropriate time for this particular assessment. While Mauritius is invariably placed within the top third of developed countries on these indicators, it rarely exhibits the very high ranking that might, deductively, be expected to form the basis of a spectacular fertility decline mediated through changes in the norms and values relating to child-bearing and parenthood. It certainly falls short of the development levels that have fostered fertility decline in Hong Kong, Singapore, Taiwan, South Korea, Cuba and Barbados.

Elaboration of, and qualification to, some of the development indicators is required. The poor performance on income (gross national product per head) cannot be rationalized, as in some less-developed countries, by a reliance on subsistence livelihoods. In fact, such is the dominance of sugar-cane cultivation on estates and small-holdings alike that a cash crop and wage economy has long been omnipotent, with 70 per cent

Table 9.2 Mauritius: ranking on selected variables among 94 less-developed countries, *c.* 1970

Variable	Mauritius value	Rank
GNP per head	US $444	33
% of employed males 15–64 in non-agricultural occupations	58 %	8
% of population urban	44 %	24
% of adults literate	61 %	28
% of population 5–19 enrolled in schools	61 %	33
Life expectancy at birth	64 years	14
Infant mortality rate	57 per 1,000	16
Population per doctor	2,700	24

Sources: Mauldin and Berelson (1978), Table 9.6 and Nortman (1976: Table 3).

of all households now holding bank accounts (Ghosh 1988). Moreover, so severe are income disparities within Mauritius, that the great bulk of population has well below the average income; in 1981, 47 per cent of the national income accrued to the top 10 per cent of households, a concentration only exceeded by Brazil in a set of twenty-six less-developed countries (World Bank 1986).

Mauritius took appreciable steps towards universal primary school education in the 1950s and 1960s, so that by 1972 the enrolment rate at primary level was 93 per cent for male children aged 5–12 and 91 per cent for females. However, the provision of secondary school education has proved difficult. There were only seven government secondary schools in 1977, and although these had expanded to twenty-four by 1985, they only accommodated 22 per cent of all secondary pupils. Private schools all charge fees, as did the government secondaries until 1977.

It is widely accepted that growing female participation in employment, and particularly in secondary school education, has been one of the major factors promoting later marriage and smaller family norms in those less-developed countries that have experienced significant fertility decline. In Mauritius, however, the female activity rate (proportion of all females over 15 years of age who are employed) was only 18 per cent in 1962 and 20 per cent in 1972. Certainly this had increased to 28 per cent in 1983 and is well above that today because of major recent investment in knitwear manufacture for export, attracted by cheap labour and political stability (van den Driesen 1988; Jones 1989a); but this recent expansion post-dates the fertility transition. Similarly, the influence of a secondary school education on female attitudes towards family size seems unlikely to have been a decisive influence in the Mauritius fertility decline. The school enrolment rate for females aged 12–17 had only reached 37 per cent in 1972 and 51 per cent in 1983, and a 1985 survey (Mauritius Ministry of Health 1987) revealed that 52 per cent of women aged 15–49 had not completed primary school and 36 per cent could not read a newspaper.

At the beginning of the major fertility decline, only very small proportions of young females had received any secondary education. The education experience of these young women themselves is likely to have had less impact on desired family size than the financial costs that would have to be borne in educating their own children as universal primary school education became established and as largely private, secondary school education expanded. Such has been the competitive nature of the Mauritian education system, based on restricted employment opportunities available to a rapidly-expanding labour force, that most pupils, even at primary level, receive fee-paying, out-of-school tuition. It seems, therefore, that the direct costs attributable to education, rather than indirect opportunity costs, are likely to have been the important ones underlying the Mauritian fertility decline. Similar findings have been advanced for Bali, another densely-populated island, where

> the concern over agriculture as a viable occupation for the next generation has coincided with a dramatic increase in emphasis on education. This is partly government development strategy and partly an almost desperate attempt by parents to ensure secure employment for their children. At this time secure employment very largely means a Civil Service job . . . what this requires is a large investment in just a few children.
>
> (Streatfield 1986:139)

Cultural background

The cultural context in Mauritius does not suggest itself as one favourable to fertility reduction. Most world religions actively or passively uphold the ideal of large families, and religion is an important part of most Mauritians' lives. In a survey of 552 adults, 58 per cent reported that they attended religious institutions regularly, 38 per cent sometimes and only 4 per cent never (Mehta 1981). About half of the population is Hindu and a further 16 per cent Muslim, both groups being descended from indentured labourers brought in to work the sugar estates in the nineteenth and early twentieth centuries. About 30 per cent of the population is Roman Catholic, comprising a very small, but economically dominant group, descended directly from the early French settlers, as well as the descendants of African slaves and the so-called Creoles of mixed African and European extraction; the 3 per cent Chinese in the population are also largely Catholic in religion.

The importance, then, of religion in Mauritius, and particularly of the major pro-natalist religions, Roman Catholicism and Islam, can be expected to deter fertility reduction. On the other hand, there is little evidence of explicit pro-natalism being practised in any segment of the community for the essentially political reasons discussed by Day (1968) and evident, for example, in contemporary Malaysia (Dwyer 1987). Although there is considerable pride in ethnic and religious identity within the plural society of Mauritius, with very little intermarriage and a significant degree of geographical segregation, intergroup relations are characterized by tolerance and the absence of intimidation and violence. It is also obvious to all that in an era of universal adult suffrage, the electoral majority of Indians, and specifically Hindus, is unassailable.

Family-planning programme

While the developmental and cultural context has not been especially favourable to fertility reduction, the strength of the particular family-planning programme adopted in Mauritius certainly has been. This strength can be demonstrated crudely in the form of input and output measures. One index of programme effort based on fifteen criteria ranked Mauritius tenth among less-developed countries in the early 1970s (Mauldin and Berelson 1978); and, in terms of programme funding per capita of national population, Mauritius always ranked in the top three in the 1970s in the annual compilations of Nortman *et al* (1969).

Three major factors seem to be responsible for the strength of the Mauritius programme: the early and widespread recognition of an acute population pressure problem, the logisitical advantages provided by the island's size, settlement pattern and infrastructure, and the particular policy emphasis adopted by the programme.

Recognition of population pressure

The post-war situation in Mauritius of severe demographic growth, population congestion and economic stagnation prompted the establishment as early as 1953 of an official Committee on Population. Its report urged government to promote family-planning services and to seek emigration outlets, but so strident was the opposition of powerful Catholic interests that the report was never debated in the legislature (Greig 1973). However, by the early 1960s a series of government-commissioned reports by prestigious external experts on economic and social policies (Meade 1961; Titmuss and Abel-Smith 1961; Hopkin 1966) all pointed to the urgent need for fertility reduction and for government programmes to achieve this.

These influential and explicit proposals prompted considerable public debate in Mauritius, persuading the Catholic Church in 1963 to withdraw its opposition to government funds being used to promote what it considered to be unacceptable family-planning methods. Indeed, in 1963 the Catholic Church fostered the establishment of an organization, Action Familiale, to teach natural methods of birth regulation. This major legitimization of a family-planning programme had taken some ten years to achieve from the first proposals in 1953, yet 1963 was still a very early date for comparable achievements among less-developed countries.

Logistical advantages

Territorially compact nations like Mauritius, less than 35 miles long and 25 miles wide, have obvious advantages for the publicizing and operation of government programmes like family planning. Communication advantages are furthered in Mauritius by the density of its road system and bus services and by the intensity of media penetration. Newspaper circulation in the mid-1970s of 90 per 1,000 population was well above the 50 per 1,000 recorded by middle-income developing countries as a whole (World Bank 1983a), and a survey indicated as many as 82 per cent of Mauritian households possessing a radio and 73 per cent a television (Mauritius Ministry of Health 1987). The same survey revealed that two-thirds of married women were aware of family-planning publicity on both radio and television.

Another important advantage is that the rural population is clustered almost entirely in hamlets and villages, facilitating accessible supply facilities and domestic visits by fieldworkers.

Policy strategy

From the outset, explicit emphasis has been given to accessing the whole population, with none of the urban bias that is present in many programmes. This has been achieved through a dense, well-located network of clinics and supply centres, and through the adoption of a labour-intensive motivational programme using low-level community workers. In 1985 only 14 per cent of the urban population and 11 per cent of the rural population had more than thirty minutes travel time to a family-planning facility (Mauritius Ministry of Health 1987). The locational pattern of family-planning facilities and personnel is examined further in Jones (1989b).

The rural emphasis of the programme may well have a political basis in that Hindus dominated the early family-planning organization and post-independence governments, and Hindus are particularly concentrated in rural areas. Certainly the 1971–5 Four-Year Plan allocated major investment to an ambitious Rural Development Programme to improve living conditions in villages by providing amenities like health centres.

The preceding discussion has suggested, largely on the basis of deductive reasoning, that fertility decline in Mauritius is likely to have been hindered rather than helped by cultural composition, assisted only modestly by ongoing development, but aided significantly by the particular type of family-planning programme intervention. An important element in both the development and programme dimensions has been identified as the wide appreciation of the costs of population growth in a congested island society. Such is the small size of Mauritius, the feeling of oceanic isolation and the extent of media penetration, that there is a significant collective demographic interest and community accountability operating at *national* level. Empirical evidence from the actual pattern of fertility decline will now be examined to further this argument.

COMPONENTS OF FERTILITY CHANGE

Table 9.3 presents the results of a decomposition of crude birth-rate decline using the method of Retherford and Cho (1973). In both intercensal periods changes in age structure have been unfavourable to crude birth-rate decline. The effect on age structure of net emigration, concentrated in the young adult groups, has been exceeded by the effect of incipient fertility decline in raising the proportion of population in the reproductive ages.

Table 9.3 Mauritius: components of crude birth-rate decline

	1962–72	*1972–83*
Crude birth-rate per 1,000:		
at beginning of period	38.5	24.8
at end of period	24.8	20.6
Decline in CBR per 1,000 due to changes in:		
age-sex structure	– 3.3	– 4.4
marital structure	7.2	1.7
marital fertility	9.8	7.3

Assumptions: All births are attributed to legally, religiously or consensually married women. The percentage of all 'married' women in consensual unions was 7.6% in 1962, 4.9% in 1972 and 3.6% in 1983, so that the decomposition results should be minimally biased by this level and variation.

Source: Age structure, age-specific fertility and marital-age-specific fertility calculated from data in 1962, 1972 and 1983 Censuses of Mauritius.

Changes in marital structure have had a particularly depressing effect on the birth-rate in 1962–72, as in several less-developed countries showing early fertility decline (Maudlin and Berelson 1978). Table 9.4 details these changes in marital structure which stem, above all, from the transition in the 1960s to a pattern of later age at first marriage among females. As elsewhere in the Third World, it is easier to identify this marriage transition than to explain it (Hein 1982). Normally-cited factors are increased levels of female education and female employment, but it has been shown earlier that in Mauritius these did not occur significantly until the 1970s. A subsidiary factor thought to have been operational in some countries, certainly in Tunisia and Sri Lanka (Duza and Baldwin 1977), has been increased economic difficulties, and this may well have been a decisive factor in Mauritius, stemming from the changing fortunes of the sugar industry.

Table 9.4 Mauritius: percentage of females currently married by age-group*

Age-group	1931	1944	1952	1962	1972	1983
15–19	30.7	35.9	39.9	27.8	12.4	10.5
20–24	62.8	65.4	72.3	68.2	49.7	47.9
25–29	74.9	74.4	82.9	83.1	76.0	70.8
30–34	76.4	76.1	83.8	85.3	83.6	76.8
35–39	75.0	74.0	81.8	83.9	84.2	79.1
40–44	68.1	67.9	76.5	78.5	80.0	78.9
45–49	63.3	61.6	69.0	71.1	74.4	75.0

* legally, religiously or consensually married women.
Source: Calculated from data in Censuses of Mauritius.

Depression in the sugar industry around 1930 was reflected in increased mortality (Figure 9.2) and in marriage postponement (Table 9.4). Conversely, a boom in production and earnings in the late 1940s and early 1950s was reflected in declining mortality (attributable also to malaria eradication) and in earlier marriage (Table 9.4). The collapse of the sugar boom and the deepening employment crisis of the 1960s may well have induced the marriage postponement in the 15–30 female age groups that is such a feature of Table 9.4. The plausibility of this theory is enhanced when marriage trends are disaggregated by ethnic group, since the 1962–72 trend towards later marriage was particularly prominent in the Indian population, which had a long tradition of early and almost universal female marriage. Significantly, the Indian population is heavily concentrated in the countryside, so much so that the two urban districts of Port Louis and Plaines Wilhems contained only 39 per cent of the island's Indian population in 1972, but 65 per cent of the rest of the population. It seems reasonable, therefore, to attribute marriage postponement in Mauritius as much to the population's appreciation of economic difficulties as to any significant growth in female employment and education.

TEMPORAL FLUCTUATIONS IN FERTILITY

Figure 9.2 and Table 9.1 suggest that there have been three major temporal trends in Mauritian fertility in recent decades:

appreciable reduction from the early 1960s to the early 1970s, stabilization in the 1972–9 period, and further decline in the 1980s. The relationship with macroeconomic conditions seems clear. The early period was one of economic difficulties, delayed marriage and fertility decline. Thus, Table 9.5 reveals that, in 1964–72, growth in the labour force well exceeded growth in employment, so that unemployment rose appreciably.

The 1972–9 period saw a marked reduction in the decline of the total fertility rate and an actual increase in the crude birth-rate. This may well have been associated with significant improvements in income and employment (Table 9.5), resulting from a boom in sugar prices and production in 1973 and 1974 and from expansionary government policy. Manufacturing was stimulated by the Export Processing Zone Act of 1970, and the 'Travail pour Tous' and Rural Development programmes recruited construction workers for rural roads, schools, health centres and village halls. Unemployment fell and, in an attempt to improve income distribution, the wages and salaries in the public sector were allowed to increase faster than the cost of living.

Table 9.5 Mauritius: economic and employment growth

	Growth per annum in GDP per head	*Growth per annum in labour force**	*Growth per annum in employment***
1964–72	1.8%	3.8%	2.1%
1972–79	8.0%	3.8%	4.2%

Notes: * Labour force includes employed and unemployed.
    ** Emloyment includes government work programmes.
Source: World Bank (1983b).

But economic growth and optimism were short-lived. Sugar prices deteriorated in the late 1970s and crop production in 1980 was badly hit by two cyclones. It had also become obvious that the government's expansionary financial policy was not sustainable; it had led to accelerated inflation, large deficits on the balance of payments, depletion of foreign exchange reserves and resort to major borrowing overseas. Consequently, International Monetary Fund (IMF) pressure in 1979 heralded a new period of stricter economic management and slower growth (World Bank 1983b). A major thrust of the new policy was to reduce consumption and pay awards, so that real incomes of public sector employees declined appreciably between the late 1970s and mid-1980s (Ghosh 1987), while unemployment rose to some 20 per cent of the labour force in 1984. It seems significant, therefore, that fertility decline was resumed in this harsher economic climate. The demographic response to renewed growth based on export-manufacturing from the mid-1980s has yet to be monitored.

The positive relationship between economic welfare and fertility, mediated through both marriage and marital fertility, that has been proposed here for Mauritius might appear to contradict the well-known secular association between socio-economic transformation, on the one hand, and fertility decline on the other. Yet the Mauritius trends, occurring in a highly-monetized economy, are entirely consistent with the basic economic theories of fertility of Becker (1960) and Easterlin (1976), in which household resources and costs figure so prominently as explanatory variables.

Easterlin argues that the critical determinant of a couple's demand for children is the relationship between their resources and their aspirations for consumption of goods and services. If their resources are scarce relative to their aspirations, they will be more reluctant to have additional children. This seems precisely what has been happening in Mauritius in recent decades. Growth of material aspirations has been fostered by the images of affluent life-styles projected by the expanding education system, by growing numbers of tourists and by appreciable media penetration. However, resources available to most families have not expanded commensurately, and, at times, have actually declined.

Freedman *et al.* (1981:15) provide similar findings from the congested island of Bali that 'contradict the conventional wisdom that families who are desperately poor will have no interest in family limitation ... sheer Malthusian pressure coupled with aspirations arising out of access to outside influences ... may increase contraceptive use among the poor'. There are also important comparisons with a much more highly developed, but similarly congested island, Singapore, where lack of natural resources, inadequate water supply, and the virtual absence of agricultural land convey to residents a sense of precariousness and an awareness of environmental limitations. The pressure of population on resources is visible to the naked eye, so to speak, and is reinforced by frequent references to this problem in the press and in government policy announcements.

(Fawcett and Khoo 1980:551)

INTERNAL VARIATIONS IN FERTILITY

There is a predictable pattern of differential fertility within Mauritius by ethnicity, urbanization and education (Table 9.6). However, the differentials are more modest than in other less-developed countries, and they have been declining as demographic homogenization proceeds rapidly.

Table 9.6 Mauritius: mean number of children ever born alive to ever-married women aged 15–49,* 1985, by selected characteristics of women.

Religion			Residence		None or incomplete primary	Education	
Hindu	Muslim	Christian	Urban	Rural		Complete primary	Secondary
3.3	3.1	3.2	3.0	3.4	3.4	3.0	2.8

Note: * Standardized to years-since-marriage distribution of whole survey population.
Source: Mauritius Ministry of Health (1987): Table 8.

It is difficult to disentangle the individual influences of ethnicity, residence, education and, indeed, other variables; there is, for example, a positive association between Hindu population, rural residence and lower education attainment. It is more important to recognize the comparative slightness of the differentials, indicating how pervasive the small-family norm has become in Mauritius. Mean ideal size of family for female respondents in a 1975 survey (Hein 1977) was 3.0; two-thirds of those with two children, and 85 per cent of those with three, said they did not want any addition to their family. Significantly, the main reason given was economic, concern being expressed about the financial burden of an additional child.

175

By 1985 the two-child family had become the ideal. Of survey respondents with two children, 85 per cent wanted no more, while an ultimate total of two children was desired by 62 per cent of women with one child and by 77 per cent of childless women (Mauritius Ministry of Health 1987). The 1975 and 1985 surveys revealed that knowledge of at least two contraceptive methods was practically universal. The only differences that did exist within the Mauritian population concerned specific methods.

CONCLUSION

Two factors have emerged from this discussion as underpinning the appreciable modern fertility decline in Mauritius. First, it does seem that in small crowded islands like Mauritius, Bali and Singapore there is often a concordance between the private costs and social costs of high fertility that may be absent in larger societies, where external diseconomies of individual behaviour can be disregarded more readily. Second, the highly-monetized nature of the Mauritian economy, stemming from the early development of a cash crop and wage economy, depresses fertility at times of economic hardship. These findings provide support, therefore, for the demographic distinctiveness of small, plantation-economy islands.

ACKNOWLEDGEMENTS

The author wishes to thank the Mauritius Ministry of Health for access to personnel and data, and the Nuffield Foundation, the Simon Population Trust and the Carnegie Trust for the Universities of Scotland for financial support.

REFERENCES

Becker, G. (1960) 'An economic analysis of fertility', in *Demographic and Economic Change in Developed Countries*, Princeton: Princeton University Press, National Bureau Committee for Economic Research.

Caldwell, J., Harrison, G. and Quiggin, P. (1980) 'The demography of micro-states', *World Development* 8: 953–67.

Cleland, J. and Singh, S. (1980) 'Islands and the demographic transition', *World Development* 8: 969–3.

Day, L. (1968) 'Natality and ethnocentrism: some relationships suggested by an analysis of Catholic-Protestant differentials', *Population Studies* 22: 27-50.

Duza, M. and Baldwin, C. (1977) *Nuptiality and Population Policy*, New York: The Population Council.

Dwyer, D. (1987) 'New population policies in Malaysia and Singapore', *Geography* 72: 248-50.

Easterlin, R. (1976) 'The conflict between aspirations and resources', *Population and Development Review* 2: 417-25.

Fawcett, J. and Khoo, S-E. (1980) 'Singapore: rapid fertility transition in a compact society', *Population and Development Review* 6(4): 549-79.

Firth, R. (1957) *We the Tikopia: A Sociological Study of Kinship in Primitive Polynesia*, 2nd edn, London: George Allen & Unwin.

Freedman, R. and Berelson, B. (1976) 'The record of family planning programs', *Studies in Family Planning*, 7(1): 1-40.

Freedman, R., Khoo, S-A., and Supraptilah, B., (1981) 'Use of modern contraceptives in Indonesia: a challenge to the conventional wisdom', *International Family Planning Perspectives*, 7(1), 3-15.

Ghosh, R. (1988) 'Macroeconomic policy and structural change in Mauritius', in *Indian Ocean Islands Development*, ed. R. Appleyard and R. Ghosh, National Centre for Development Studies, Australian National University, Canberra.

Greig, J. (1973) 'Mauritius: religion and population pressure', in *The Politics of Family Planning in the Third World*, ed. T.E. Smith, Allen & Unwin, London.

Hein, C. (1977) 'Family planning in Mauritius: a national survey', *Studies in Family Planning*, 8(12): 316-20.

Hein, C. (1982) *Factory Employment, Marriage and Fertility: The Case of Mauritian Women*, World Employment Programme Research, Working Paper 118, International Labour Office, Geneva.

Hopkin, W.B. (1966) *Policy for Economic Development in Mauritius: Objectives and Principles*, Government of Mauritius Sessional Paper 6.

Jones, H. (1989a) 'Mauritius: the latest pyjama republic', *Geography* 74: 268-9.

Jones, H. (1989b) 'Fertility decline in Mauritius: the role of Malthusian population pressure', *Geoforum* 20: 315-27.

Lamusse, R. (1980) 'Labour policy in the plantation islands', *World Development* 8: 1035-50.

Mauldin, W.P. and Berelson, B. (1978) 'Conditions of fertility decline in developing countries, 1965-75', *Studies in Family Planning* 9(5): 89-147.

Mauritius Ministry of Health (1987) *Mauritius Contraceptive Prevalence Survey 1985: Final Report*.

Meade, J.E. (1961) *The Economic and Social Structure of Mauritius*, London: Cass.

Mehta, S. (1981) *Social Development in Mauritius*, New Delhi: Wiley.

Nortman, D. (1969 and following years) *Population and Family Planning Programs: a Factbook*, New York: The Population Council.

Retherford, R. and Cho, L. (1973) 'Comparative analysis of recent fertility trends in East Asia', in *Proceedings of the International Population Conference*, 2, Liège: International Union for the Scientific Study of Population.

Sombo, N. and Tabutin, D. (1985) 'Tendances et causes de la mortalité à Maurice depuis 1940', *Population* 40: 435–53.

Streatfield, K. (1986) *Fertility Decline in a Traditional Society: The Case of Bali*, Canberra: Australian National University, Indonesian Population Monograph Series.

Titmuss, R. and Abel-Smith, B. (1961) *Social Policies and Population Growth in Mauritius*, London: Cass.

Tsui, A. and Bogue, D.(1978) 'Declining world fertility: trends, causes, implications', *Population Bulletin* 33(4).

van den Driesen, I. (1988) 'Industry', in R. Appleyard and R. Ghosh (eds) *Indian Ocean Islands Development*, Canberra: Australian National University, National Centre for Development Studies.

World Bank (1983a) *World Tables*, 2, Baltimore: Johns Hopkins University Press.

World Bank (1983b) *Mauritius: Economic Memorandum: Recent Development and Prospects*, Washington, DC: A World Bank Country Study.

World Bank, (1986) *World Development Report 1986*, New York: Oxford University Press.

10

LAND SETTLEMENT SCHEMES IN JAMAICA AND FIJI

Progress through transformation

Michael Sofer and Israel Drori

INTRODUCTION

During the last four decades Land Settlement Schemes (LSS) have been a research focus for social scientists of different disciplines, especially for scholars occupied with issues of rural development. A considerable portion of the concerned literature deals with the evaluation of the performance of settlement projects as a strategy for rural development. Historically, LSS were already initiated in colonial times, but their popularity has been retained under independent governments, and they are currently being promoted in a substantial number of developing countries (Maos 1984; Oberai 1986; Hulme 1987). The main objectives of this strategy include improving welfare levels, redistribution of land and population, assisting the poorest and landless sections of the population, and improving farmers performance through an increase in food production. In most cases reviewed, however, LSS performance has not been regarded as a major success (Hulme 1987), a finding which seems to contradict the continuing support given to land settlement initiatives by national administrations and international land agencies.

The term 'land settlement scheme', although widely used in the literature, is not precisely defined and needs some clarification. LSS appear in a variety of forms based in different ideologies. Schemes can be of a small-holder form or co-operative in nature, and their population may occupy new lands or their own newly-redistributed lands. They may be state-sponsored projects or may be stimulated by farmers' spontaneous initiative. In this paper we regard 'land settlement schemes' as 'projects in which a group of people moves permanently or semi-permanently to occupy an area of unused or underutilized (in a market economic sense) rural land, under the guidance of an agency external to the settler community' (Hulme 1988:42). These rural projects are regarded as a part of an integration mechanism that links together the production, organizational and technological spheres (Drori 1990).

The purpose of this chapter is to review certain socio-economic changes which occurred in rural committees in the island states of Fiji and Jamaica, as a result of the introduction of state-sponsored LSS. The main argument of this chapter, which receives support from African experience (Schaaf and Manshard 1989), is that the transformation process is a two-stage process. The first stage is an institution-based process, and the second stage is an adaption process largely based on a spontaneous mechanism initiated by the settlers, as a result of inflexibility within the planned institutional process. The handover of the scheme to the settlers, with the implementing agency acting as a side supporter, has had a favourable influence on the scheme's performance. Therefore, the shift from the first to the second stage appears to be a pre-condition for an adequate adaptation which may lead to success.

SMALL ISLAND STATES AND LAND SETTLEMENT SCHEMES

Historically, small island states were characterized by their agricultural economies. Even though the situation has changed since colonial times and other sectors, mainly tourism, have become dominant, agriculture remains the most important sector in terms of employment. In colonial, and especially in tropical island states, agriculture was characterized by two sectors operating side by side; the first sector was export-oriented and based largely on plantations; the second was that of peasant production which came increasingly under pressure through loss of land and population growth (see, for example, Brookfield 1972a; Beckford 1975b; Levitt and Best 1975). Labour linked the two sectors in a form of dual dependence; plantations depended on cheap labour, and villagers sought an important source for additional income through temporary wages on plantations.

The consequences of the colonial development pattern became manifest at independence by problems ensuing from the dominance of monoculture, the marginalization of native production, decreasing returns from village agriculture, environmental degradation, and the increasing peripheral position on world markets (Demas 1965; Benedict 1967; Sofer 1988). Moreover, in most island states rates of increase of food production did not keep pace with demographic growth (Connell 1988:36), thus posing a potential situation for food dependency.

Failure of the agricultural sector to respond to the increasing demand for food may be attributed to four main factors: first, skewed capital allocation where the heavy concentration of productive investment was in a very small number of export crops, such as Fiji and Jamaica; second, generally low investment in village agriculture especially in native (village) food production (Sofer 1987a); third, rapid population growth in urban areas, partly based on out-migration from rural areas which, in turn, might have contributed to a decline in farm hands (Connell 1981); and finally, land tenure problems, such as small land area or communal ownership, which has limited farmers' access to credit (Sofer 1987b).

In the framework of such an economic structure and its internal constraints, a number of island governments promoted agricultural development policies based on LSS. In island states where unutilized or underutilized land is still available, settlement projects may provide land to the landless, relieve overpopulation on limited arable lands, solve economic and social problems subsequent to natural disasters, and expand food production. Furthermore, the underlying assumption is that the imposed transformation of the production form inherent in the LSS strategy implies a more efficient use of land and labour resources, as well as capital resources, where available. In addition, state-sponsored projects may attract private capital, while state investments in infrastructure and credit facilities may stimulate the owners of private capital to initiate local ventures. This strategy was applied in larger island states such as Papua New Guinea (Hulme 1984) and Sri Lanka (Dunham 1982) and also in smaller island states such as Fiji (Brookfield 1979; Sofer 1987b) and Jamaica (Drori 1990).

The conceptual framework discussed above is the background for the presentation of two small island states case studies, where Land Settlement Schemes strategy was applied. Two different experiences, in Fiji and Jamaica, may serve to demonstrate the settlers' adaptation to state-initiated projects.[1]

FIJI

The Fijian case study shows the outcome of substantial modifications which occurred in the production system of Narata's village community. These modifications constituted a fundamental transformation and created the necessary conditions for this rural community to break away from the marginalized situation of a village community based on mixed subsistence-cash cropping. Principally, the transformation meant that the shift from communalism to individualism led to a more intensive use of production factors, greater commercialization, and higher levels of income.

Narata village and the land settlement scheme

Narata village is located in the Singatoka Valley on the drier side of Viti Levu (Figure 10.1) The valley is occupied by both Fijian villagers and Indian small-holders, growing a variety of food crops, especially root crops, vegetables, corn and fruits, together with industrial crops such as sugar, passionfruit and tobacco. The village consists of four *matangali*[2] and twenty-four households which in 1982 comprised 128 persons (excluding six temporary absentees); 65 females and 63 females. Regarding the mobility of villagers in pursuit of outside wage opportunities, it was recorded that three absentees were employed outside the village as wage workers, two in non-temporary urban jobs, and one in a temporary rural job. Of the twenty-four households, eight resided on their land; three divided their time between their houses inside and outside the village; and thirteen household members domiciled in the village.

Narata village was part of the *Dumbalevu Development Scheme* which was introduced in 1965 as one of five schemes established in the Nandroga/Navosa province[3] during the early 1960s, four of which were established in the Singatoka Valley. The schemes were introduced by the Land Development Authority (LDA) established in 1961 when a new approach to

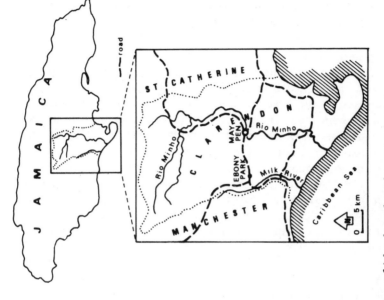

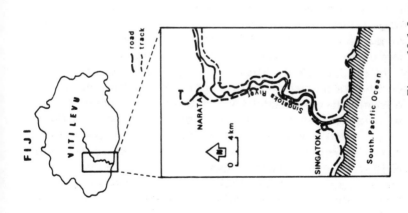

Figure 10.1 Location map of islands and settlement schemes

rural development was needed. The LDA considered four main obstacles confronting adequate rural development, namely, lack of security of tenure, lack of productive capital, inadequate infrastructure including roads, assured markets and technical advice, and lack of productive incentives on the part of the villagers (LDA 1963:3). Its main declared tasks were to open up new lands not settled or leased, and the promotion of increased production and new crops, where possible, on land held by existing smallholders (LDA 1963:1). The strategy chosen as the best possible response to these obstacles was settlement schemes supported by investments in roads and land clearance, organization of marketing services, allocation of redistributed land and security of title, provision of credit for farm equipment and professional supervision of new crops.

The history of the scheme recorded fluctuations in farmers' interest and success. Most of Narata's villagers moved to their allocated blocks of land with the initiation of the scheme and followed the cropping programmes instructed by the scheme's supervisors (Sofer 1987b). Towards the end of the 1960s the scheme faced some serious difficulties, due mainly to lack of credit and support services, difficulties in following the rigid cropping programmes and farmers' indebtedness. Consequently, some of the farmers abandoned their land and moved their residence back to the village. Although the landownership pattern established by the LSS remained the same, this shift meant a relapse to past cultivation patterns. It was associated with an increase in the proportion of subsistence farming, using proportionately less of the total acreage, and limited advanced technology.

In the late 1970s, after more than ten years of relative decline, local extension officers of the Ministry of Agriculture operating in the Singatoka Valley tried to encourage the old schemes' farmers to extend their commercial production (MAF 1980), a move which met with enthusiastic response from Narata's farmers. The officials did not repeat past patterns of rigid cropping programmes and patronizing supervision. The main support was through facilitating access to loans, while enabling the farmers to engage in cultivation in response to local and regional market conditions. In 1979, loans became available through the Fiji Development Bank and were primarily allocated for land penetration and the purchase of farm inputs and capital items (MAF 1980). By early 1983 Narata's farmers managed to achieve a level of technology and income higher than what was generally common in the native Fijian rural community.

The production pattern in Narata after the initiation of the LSS

It is worthwhile to look at a number of socio-economic indicators which illustrate the villagers conditions after the initiation of the LSS. The main emphasis was on changes in the use of labour resources, land distribution, pattern of agricultural production, use of means of capital and income pattern. Some of the data can be compared with villages which have not experienced the scheme (Sofer 1987a). Some data are also compared with data collected by Belshaw in 1958 during his study of the Singatoka Valley (Belshaw 1964), thus enabling a comparison with the performance of the village production system before and after initiation of the land settlement scheme.

Use of labour resources

The majority of Narata's labour force is engaged in farming activities. Excluding the temporary absentees, no wage-earners were recorded in the 1982 fieldwork. The number of full-time farmers, who were exclusively men, was thirty four. Yet most other villagers, either men or women of fifteen years of age and over, were recorded as part-time farmers. In addition, nine persons (eight men who were also enumerated as farmers and one woman) were recorded as middlemen who were largely engaged in market vending in the west coast towns.

It is common for a Fijian villager to divide his time between three types of activities: farming, communal obligations and leisure. Although there is no clear division which may suggest the proportional share of each of the above activities, it seems that the tendency for an increase in the proportion of working hours in farming activities is positively correlated with the proportion of market-orientated production of the villager. Table 10.1 shows the average working time per week devoted to farming activities of five Fijian villages in different parts of the country. The other four villages were not subjected to any LSS initiative. It should be pointed out that some variables may have a considerable influence on the length of working time devoted to agriculture, such as the physical environment, the size of area under cultivation, and the type of crops grown. While it is clear that the first variable affects the other two, all these variables are also closely related to the prevailing production system. The highest average numbers of hours per week for both full-time and part-time farmers, as well as a high average of part-time farmers per household, are recorded regarding the village of Narata. The disparities summarized in Table 10.1 in the figures for average farming hours per household, are not basically generated by differences in the structure and size of the labour force. Farming time is part of the nature of each of the village production systems, in which commercial farming attitudes, as in Narata, result in a larger number of working hours on the fields as compared to less commercial farming systems in the other villages. Moreover, in contrast to Narata, in areas where the traditional social structure is strong and where communal obligation and leisure time are important, the proportion of time spent on farming is relatively low.

Table 10.1 Average number of hours per week spent in farming activities

Village	Narata	Dravu-walu	Nalotu	Votua	Naqali
Number of households sampled[1]	23	26	23	19	24
Characteristics					
Average H/W for full-time farmer	27.0	25.8	22.2	19.6	22.9
Average no. of full-time farmers per household	1.5	1.4	1.9	0.5	1.8
Average H/W for part-time farmer	12.1	5.9	6.9	7.0	10.7
Average no. of part-time farmers per household	1.4	0.65	0.65	1.9	0.9
Average H/W for household	56.8	39.6	46.0	23.6	49.3
Mode	–	36	36	44,20,11	30
Median	55	35	42	20	45.5
Standard deviation	26.8	17.5	18.0	15.8	19.5

Source: Fieldwork.
Notes: H/W = hours per week spent in farming activities.
1 Excluding households in which farming is not their major engagement.

Land distribution

Generally, before the introduction of the settlement scheme, villagers cultivated their plots within their *matangali* land. In addition, it was common for villagers to have access to land owned by other *matangali*, especially when they faced problems of scarcity of land, or plots distant from the village. As a result of the redistribution of land, the average size of the holdings was nearly nine acres (Sofer 1987b), an area much larger than the 1978 average size of 1.75 acre for the village farm in the Western Division (Rothfield and Kumar 1980).[4] The redistribution of land also helped to reduce inequality in holdings. The Gini index of inequality shows an index value of 0.19, which is low on a 0 to 1 scale. Furthermore, this figure is much lower than the corresponding figure calculated from the 1978 Agricultural Census (ibid.), which is 0.56 for the total village farms in the Western Division. The land distribution pattern manifests that the settlement transformation process resulted both in more land being cultivated and in greater equity.

Agricultural production

A comparison of crops for the two periods before and after the project was introduced is presented in Table 10.2. The cash crops that appear in Part B of Table 10.2 are also used to a certain extent as subsistence crops, though they do not always appear in Part A. By 1982 few new crops had appeared in Narata's fields, yet the frequency of the cropping changed significantly. There was a substantial increase in farmers engagement in all crops for subsistence purposes other than yams. In relation to cash crops there was a clear increase in commercialization of all major crops, other than bananas and peanuts. This increase in commercialization was mainly in the traditional root crops sector including cassava and kumala, but also in the industrial crop sector, including tobacco and passion fruit.

Although production of modern crops in Narata is partially a legacy of the period when the Land Development Authority was involved in the scheme, the revived interest in these crops represented a clear response to the renewal of capital injections in the form of loans to local farmers. After a period of concentration on crops which demanded less capital inputs (Sofer 1987b), the support of the public sector, and the involvement of agro-industrial firms providing inputs and

Table 10.2 Number and percentage of households growing
major domestic and cash crops, 1958 and 1982

Households growing:	1958[a]		1982[b]	
	No.	%	No.	%
A. Domestic crops				
traditional				
Cassava	13	100	23	100
Dalo	2	15	10	43
Yams	13	100	20	87
Kumala	1	8	11	48
Vundi (Plantains)	2	15	15	65
modern				
Rice	1	8	7	30
Bananas	7	54	14	61
B. Cash Crops				
traditional				
Cassava	12	92	23	100
Dalo-ni-tana	–	–	2	9
Yams	–	–	2	9
Kumala	–	–	8	35
Yaqona	1	8	1	4
Vundi	1	8	1	4
modern				
Corn	3	23	16	70
Rice	1	8	2	9
Water Melon	2	15	6	26
Vegetables	3	23	11	48
Tobacco	1	8	11	48
Passion fruit	–	–	11	48
Bananas	11	85	4	17
Peanuts	3	23	–	–

Sources: 1958 – Belshaw (1964); 1982 – Fieldwork.
Notes: a Total enumerated households = 13.
　　　 b Total enumerated households = 23. One of the 24 households was
　　　　 omitted due to the fact that the composition of its cash crop was
　　　　 not clear enough.

marketing services, enabled farmers to grow industrial crops
such as tobacco and passion fruit.

Capital resources

Generally, three levels of technology may be distinguished in
Fiji's agriculture. The first level consists of manual labour using

simple agricultural hand tools for farm work. At the second level draught animals such as bullocks and horses are used. In spite of the fact that mechanization in agriculture has been on the increase in recent years, mainly in the sugar cane area, the overall level of mechanization in Fiji remains low (Chandra 1983), and its diffusion around the country is limited and selective.

Inter-village disparities in farm capital are shown in Table 10.3. Above the level of hand tools the disparity is enormous. Animal-drawn implements are used almost exclusively in Narata in most of its households. Furthermore, Narata is the only village to possess tractors – two in number – and, although they are privately owned, they are at times hired by other villagers. The use of fertilizers and pesticides is directly linked to the cultivation of industrial crops such as tobacco and passion fruit, and points to the link between the village and commercial enterprizes outside the village. Importance should also be given to the financial loans taken by villagers and to the main purpose of these loans. It is obvious that Narata's farmers have been using this capital resource much further than any other village, mainly due to the fact that they have a lease to their land, which most other villagers have not.[5] Moreover, as Table 10.3 indicates, Narata's farmers seem far more interested in productive capital than farmers in any other village. The purpose of the loans is another indicator of the productive and commercial attitudes of these farmers.

Income pattern[6]

There is only a small degree of change in the distribution of major sources of cash income after the initiation of the scheme as shown by Table 10.4. Crops still remain the principal source of cash income although their share has declined somewhat. This is partially compensated by the income increase for the category of middlemen, which is the income derived from selling crops for other farmers, either from the village or from neighbouring villages and farms. A change in the income pattern is due to the addition of new categories, especially wage income and hiring of capital items (included in 'other'), although together these contributed less that 4 per cent to the total village income in 1982.

Table 10.3 Availability of capital items, agricultural inputs and loans taken by village

Village	Narata		Dravuwalu		Nalotu		Votua		Nagali	
No. of households (1)	24		26		23		19		24	
Capital items (2)	No.	RT	No.	RT	No.	RT	No.	RT	No.	RT
Level I										
Bush knife	69	2.9	49	1.9	62	2.7	48	2.5	98	4.1
Spade	18	0.8	34	1.3	43	1.9	23	1.2	37	1.5
Fork	2	0.1	9	0.3	2	0.1	14	0.7	44	1.8
Level II										
Animal-drawn plough	21	0.9	0	0	0	0	0	0	1	0
Animal-drawn harrow	21	0.9	0	0	0	0	0	0	1	0
Level III										
Tractor	2		0		0		0		0	
Others										
Punt	0		3		3		3		0	
Boat	0		2		0		2		0	

	(1)	(2)	(3)	(4)	(5)
Agricultural inputs (3)					
Fertilizers	17 (RT 0.7)	0	0		17 (RT 0.7)
Pesticides/herbicides	17 (RT 0.7)	0	0		10 (RT 0.4)
Characteristics of loans					
(a) Number of loans (4)	12	3	1	5	2
(b) Total value of loans (F$)	17,045	1,930	600	9,060	130
(c) Major purposes for having a loan	bullocks and related implements; housing construction	family needs; purchasing goods for shop	farm inputs	boat; housing construction	farm inputs

RT = ratio, which means the number of implements per household.

Source: Fieldwork.

Notes: 1 Number of households sampled.
2 Ratios are used only where they are meaningful.
3 These numbers represent the number of farming households using such inputs, and the ratio represents the proportion of this number of total village households.
4 Votua had another loan for house construction, but the amount of this loan was not disclosed.

Table 10.4 Distribution of sources of cash income for Narata village, 1958 and 1982 (in F$ [2] and per cent)

	1958[3] Total ($)	%	1982 Total ($)	%
(a) Crops[4]	15,632	76.2	4,3201	70.3
– Cassava, root crops, vundi	6,922	44.3	26,319	63.0
– Bananas	7,140	45.7	558	1.3
– Modern crops[5]	1,570	10.0	10,510	25.2
– Industrial crops[6]	–	–	4,314	10.3
(b) Livestock and fish	–		1,154	1.9
(c) Middlemen	4,452	21.7	14,860	24.2
(d) Rent and remittances	420	2.1	20	0.03
(e) Wages[7]	–	–	950	1.5
(f) Hiring out bullocks & tractors	–	–	1,245	20
Total	20,504	100.0	61,430	99.9
Cash income per household[8] – mean	1,599.2		2,559.6	
– median	1,046		2,374	
– standard deviation	1,304.5		1,664.6	
– coefficient of variation	0.82		0.65	

Sources: 1958 – Belshaw 1964: 206–7; 1982 – Fieldwork.

Notes
1 Number of sampled households: 1958 – 13; 1982 – 24.
2 At the time of the survey the value of one Fijian dollar was about 1.1 Australian dollar.
3 1958 figures are in real prices of 1982. The conversion factor is: 1F $ = 2F $, and this is multiplied by 4.2 which is the 420% increase in the Consumer Price Index (CPI) since 1958.
4 The internal subdivision is the percentage share of the sum of F$ 41,701 which excludes one household which could not divide its income according to different crops.
5 Modern crops include: corn, rice, vegetables and water-melon.
6 Industrial crops include tobacco and passion fruit.
7 Wages of villagers working temporarily in cutting sugar-cane and for The Pine Commission.
8 The 1958 statistics are based on mid-class values of the income grouping in Belshaw which brings the village total income to F $20,790; 1.4% higher than the total of F $20,504 presented here.

The major changes, however, appear within the types of crops grown for cash. Income from cassava production increased as production was targeted for the urban markets and marketed by the farmers themselves. Conversely, the banana crop was discontinued as a major source of cash income due to the 'black leaf' disease. This was compensated mainly by the introduction of modern and industrial crops and an increased contribution of cassava.

Between 1958 and 1982 cash income per household increased on an average, in real terms, by 60 per cent, derived from a combination of a larger share of commercial production (at the expense of subsistence production) and from an increase in productivity. In addition, it seems that cash income became more evenly distributed among households, as seen by the values of the median and the coefficient of variation in Table 10.4. Both the rise in real incomes and their more even distribution, which has essentially followed agricultural changes, are significant developments.

JAMAICA

The Jamaican case shows the transformation process of a settlement scheme in the context of co-operative development. Principally, this transformation meant that the shift from collectivism to a so-called 'group-farming' organization of production achieved more effective farming in terms of use of labour, land and new technology. This has resulted in higher levels of production and income.

The scheme and the local agricultural setting

The scheme known as Ebony Park, located in the sugar belt of mid-Clarendon Plain (Figure 10.1), consists of two villages comprising 210 families. The scheme was under the administrative responsibility of the Social Development Commission which was the implementing agency of the Ministry of Sport, Youth and Community Development. It was administered by a professional management, mainly a project manager who was in charge of the overall developments of the scheme, and a farm manager who concentrated on the agricultural aspect. The scheme was part of a number of agrarian schemes in mid-Clarendon introduced in the 1970s, which also included Pioneer Farms, Sugar Co-operatives and Project Land Lease (NPA 1978).

The scheme was established on land which was formerly under pasture for beef cattle and was also planted with citrus and sugar cane. The Jamaican Government bought the land from its private owners and initiated the scheme in 1978, based on the ideology of the Peoples National Party (PNP). The PNP emphasized its commitment to a participatory form of social and economic organization and control within the economy (Manley 1974). The scheme operated as a multi-purpose co-operative, providing its members with professional management, inputs, credit, marketing services, modern technology, and access to information and technological innovations (Drori 1982 and 1990; Gayle and Drori 1984).

The scheme's main objectives were: to resettle 210 landless young families in two nuclear villages; to relieve underemployment and unemployment among young people in the area; to raise production levels through the introduction of planned and advanced agriculture; to demonstrate to local farmers the advantage of the co-operative as a vehicle for social and economic development (Drori 1982).

The co-operative was initiated as an alternative production system to the commercially-inefficient small farmer. Generally, the dominant agricultural production pattern in the area is based upon mixed farming crops. This term applies to farming systems mainly concerned with food crops, semi-permanent crops such as bananas, and permanent crops such as fruit trees, on relatively small pieces of land. The main cash crops cultivated by the small farmers are: tomatoes, cucumbers, string beans, skra, *gungo* and red peas, corn and pumpkins. Livestock is an integral part of mixed farming and most households rear either pigs or goats, as well as cattle and chickens. The well-to-do farmers in the area use a mixed farming strategy on an entirely different scale: in addition to sugar-cane, they also concentrate on poultry, beef cattle, dairy farming, citrus and pisciculture. Due to the fact that many of the area residents are labourers in the sugar industry, it is common in the off-period season between August and December that non-sugar farming obtains more attention and more acreage is planted.

The agricultural land in the area is very unevenly distributed. Seventy-eight per cent of farms, on areas up to five acres, occupy 15 per cent of the land area, while 1.6 per cent of the farms, mostly sugar estates, own 65 per cent of all cultivated land. As a result of this land ownership pattern and the existence of a large number of landless youth, the scheme's 2,100 acres were distributed differently. It was planned that approximately 1,400 acres of land suitable for intensive agricultural production would be parcelled out into five-acre plots, to be allocated to landless settlers. The rest of the land was designated for the construction of villages and communal agricultural activities.

Farmers in the area have experienced a number of constraints regarding effective commercial production related to land preparation, farm supplies, marketing and availability of labour. Regarding land preparation there is a shortage of tractors mainly in the sugar season from January to August. Consequently, farmers without access to tractors are able to prepare only a small part of their land which is done by manual labour. Moreover, during the sugar season, prices for hiring of tractors are considerably higher and the small farmers usually do not have the available cash.

As for farm supplies, the shortage of inputs such as fertilizers, herbicides, seeds and tools, is a well-known phenomenon in the area. They are either not available, their price is high, or the farmer needs to travel frequently to the nearby towns of May Pen or Mandeville to obtain them. Two main marketing channels are available to small farmers in the area: first, institutional channels such as supermarkets, schools, hospitals, restaurant and the Agricultural Marketing Co-operation (AMC); and second, non-institutional channels, mainly private intermediaries known as 'higglers'. In the small rural communities of the mid-Clarendon area there is another marketing channel known as 'selling locally', characterized by small quantities of foodstuff sold to neighbours. The small farmers in the area prefer to sell to local institutions, but only a few are able to guarantee a steady supply and thus most of them have no access to these outlets. The majority try to bargain between the AMC and the higglers and show a distinct favouritism towards the latter. The higgler has the advantage of buying the products at the farm gate, in many cases he reaps the crops in the field and usually pays better prices than the AMC.

Although agricultural production is the main activity in the area, the availability of farm labour is a persistent constraint. Labour shortages are a direct response to low wages, which are less than the minumum wages paid to casual workers in public works, and to the stigma attached to work in agriculture (Drori and Gayle 1990). Due to the fact that men are less responsive to the farmers' recruiting efforts, the rural force consists mainly of women, children and the aged. Otherwise, local farmers tend to recruit male labour force from outside the area, either from the town of May Pen or the hilly hinterland. A similar pattern occurs in the sugar industry, where cane cutters receive a scant reward for their labour. The employment situation has increased the tendency to depend upon family labour, as well as various forms of labour exchange. By an arrangement known as 'morning sport' a farmer invites five to ten persons to help him undertake a large project which has to be completed within a short period of time. In another 'day for day' arrangement, labour is exchanged. In 'partnership', farmers agree to work together (Drori 1982).

The pattern of production within the settlement scheme

During the first period under review, agricultural production was mainly carried out on some 90 acres of the scheme land. This land was irrigated from the nearby Milk River which facilitated intensive cultivation of cash crops, mainly pumpkins, cucumbers, tomatoes and green peppers. Cultivation was carried out by fifty-one settlers, the first and the second groups of settlers recruited by the scheme.

Organization of labour

Within the co-operative there were two major types of labour organizations operating on the following principles (Drori 1982:116).

First, was the *collective*. This was the original form of organization within the co-operative. In this arrangement, the settlers collectively cultivated the co-operative land. Although divided into task groups headed by group leaders, it was actually the farm manager who assumed overall responsibility for decision-making and for drawing up the farm plan. Land preparation, the supply of inputs, training, and marketing were administered by the co-operative management. The revenue from the crop went to the co-operative for recurrent expenditure and the settlers received a fixed allowance.

The second was the *group* structure. In this each settler works in one of six groups, each of which cultivates co-operatively its portion of 'the 90-acre block'. Land preparation, supply of inputs, credit and marketing for both the groups and individuals are carried out by the co-operative. Decision-making concerning production is done by the group in consultation with the farm manager, who has the last word on land utilization matters. The group activity is co-ordinated by the group leader who also represents its interests in dealing with the co-operative management. The revenue from the crops, after deducting the cost of inputs, is divided among the group members according to the amount of work invested, and after deducting the cost of inputs. The settlers continue to receive a fixed allowance.

The changes in the organization of labour are manifest through the shift from the collective to group farming. These changes expressed the settlers' spontaneous response to the unsatisfying planned organizational and economic features of the scheme as implemented by the Social Development Commission.

Two main points are the basis for the underlying changes: first, the settlers have manipulated and changed the introduced institutional framework by forming new labour organizations which correspond to their 'real world' experience. Second, although the co-operative supplies means of production, including a market outlet and guaranteed prices, this alone is not sufficient. A vital input not provided for the settlers is labour, which is crucial for intensive cultivation, increased yields and higher income. The shortage of available agriculture labour in the sugar belt area emphasizes the importance of voluntary co-operation as a substitute for hired labour.

The emergence of the *group* resulted mainly from the deep frustration felt by the settlers, and the urge to benefit more directly from their own labour. The decision to dissolve the *collective* type and to establish the group system was made spontaneously and unilaterally by the settlers themselves. They decided to use the group as the basic structure, assigning to each group an equal share of the co-operative land under cultivation. The co-operative management avoided direct confrontation with the settlers and recognized *de facto* that settlers had the right to benefit from their labour even though the land, mechanization and irrigation were supplied by the co-operative.

In the beginning the *group* system suffered from various drawbacks, mainly lack of cohesiveness, negative attitudes toward work, and friction among the settlers regarding the sharing of revenue. With the establishment within the groups of social control mechanisms such as the application of group pressure on members not contributing their full share of work, and keeping records on the amount or work that each settler contributed, production and revenue have changed for the better.

Summing up, the shift from collective to group farming has brought about first, increased production, with work being carried out quicker and with maximum land utilization, and second, greater borrowing opportunities since it became easier for the settlers, as a group, to borrow money from the various lending agencies. The group's combined initial capital was relatively high and enabled the settlers to engage fully in commercial agriculture. There is also a better supply of services; as the settlement was formed as a co-operative, the means of production were supplied by the co-operative according to the settlers' demands. As a group, the settlers' access to tractor and other farm supplies was simplified.

Characteristics of production

As already pointed out, there was a substantial increase in production level after the change to the *group* system. This increase can be attributed to both individual incentives in terms of outcome, and a secure market for the export of some cash crops, mainly pumpkins. Under the *collective* system of labour organization the first harvest of pumpkins yielded an average of 5,929 pounds per acre; six months later when the *group* system was consolidated, the average yield per acre rose to 11,100 pounds per acre. In order to get higher yields, the settlers working in groups improved their farming practices by working harder, intensive use of manure, and by doubling the number of plants per acre. The improved practices enabled more rational use of tractors, thus saving labour hours which were invested in weeding the plots more thoroughly.

The average time spent in the field grew to eight working hours per week during the 'collective' period, to thirty-eight hours per week during the *group* system. The change in the pattern of labour organization also increased the acreage under cultivation from thirty four acres in the *collective* system to sixty two acres in the *group* system. There was a notable increase in the acreage of pumpkins (from 5 to 30 acres) due also to the available export market. In addition, new cash crops such as red peas, string beans, okra and plantain, were introduced.

As for the settlers' income, at the beginning of the project the settlers received a monthly allowance of J$160,[7] which included the take-home pay ($100), compulsory savings ($20), and subsidized lunch ($40). Since the settlers began to produce for themselves, their average income has tripled. The most successful settlers in terms of income were those who engaged in both *group* and *individual* farming. Their average income as registered for a six-month period, was J$2,525 and was equivalent to the average annual agricultural income of small farmers in the area.

Mechanization

The use of agricultural machinery is presented in Table 10.5. The transformation from collective to group farming was accompanied by an increase in the use of tractors and their related accessories. In contrast to the more rational use of tractors with the *group* system, tasks such as ploughing, disc-harrowing and bush cutting were not carried out by the settlers during the collective system. This was also true in the case of local small farmers who used tractors for ploughing only. Within the *group* system, as productivity became a crucial element, and as 'time costs money', the settlers began to make fuller use of mechanization.

Table 10.5 Use of mechanization by type of labour organization (average hours/week)

	Collective	Group	Individual farming[1]
Tractor for transport	5.5	12.5	7.5
Tractor for ploughing	0.5	1.5	0.5
Disc-harrowing	–	2.5	1
Bush cutting	–	0.5	–

Source: Fieldwork.
Note: 1 *n* = 36.

CONCLUSIONS

The case studies under consideration in Fiji and Jamaica have a great deal in common despite different circumstances. The similarity is in the process of change. Both cases started as state-initiated projects, and were generally regarded by the settlers as unsatisfactory. Nevertheless, after the scheme was transformed and based on different principles which better suited both local conditions and settlers' expectations, the scheme reached a relative degree of success.

The implementation of the projects did not necessarily follow the model which the planners had in mind when the project was started. The outcome is an adjustment between the planners' intentions and local dynamics. The resulting form of organization and operation is part of a two-staged evolutionary process, in which the premilinary stage constitutes a pre-condition for the following stage of adaptation.

In the first stage, the organization of production is transformed in the spheres of land and labour use, while the second stage is characterized by a form of labour organization which has been adapted to the settlers' needs and capabilities. This is an adjustment between the traditional organization of production and the challenges raised by more intensive commercial farming. Both cases emphasize the conversion from communal to less communal forms of production; a conversion which received state support even though it was carried out by the settlers themselves. This is further characterized by increased settler participation in the revival of the project after a period of crisis and relative stagnation.

State intervention was necessary in Fiji in order to achieve the goal of modifying the subsistence-oriented (peasant) production system, and for establishing a more participatory production system in Jamaica. It can be argued that the modification in the village and co-operative production systems were not merely a spontaneous response to existing market forces, but that they happened through, and were part of, a selective policy generated and implemented by state agencies. The initial transformation, which at a later stage was based on the settlers' spontaneous adaptation, is currently sustained by these agencies by means of productive capital and supportive infrastructure.

Is all that relevant to small island states? The described situation stands out against the general trend of failures so common in land settlement schemes in developing countries. Two points may be raised here. First, that small projects may possess advantages of small size as against economies of scale. Second, the range of alternatives in small island states' agriculture is very limited. The difficulties encountered in engaging in capital-intensive crops, such as sugar in Fiji and Jamaica, is forcing farmers to look for local cash crops markets. Moreover, product concentration made diversification strategies increasingly attractive and enabled farmers to move away from dependency on a single export crop so typical of tropical small island states.

The stage of adaptation experienced by the settlers includes two related strategies: risk aversion and risk minimalization. These two strategies are normal practices of farmers in tropical island states (Brookfield 1972b). In spite of the modernization objectives put forward by the planning authorities, the settlers retained several characteristics and practices to which they were accustomed, and which were better adjusted to the local and economic environment.

The emerging pattern of farming is a compromise between new forms of production on one hand, and of local modes of production and development constraints of small island states on the other hand. A lesson to be gleaned is that a compromise between planned settlement schemes and settlers' spontaneous motivation, combining the advantages of both may serve as an adequate means for implementation of land settlement schemes in those small states. Revoking the institutional features which proved to be economically and socially inadequate, and stimulating individual economic decisions, along with a certain amount of government support, enabled the small farmers to better utilize their land and labour resources.

NOTES

1 This chapter is based on fieldwork carried out in Fiji by M. Sofer during 1981–3 and in 1988, and in Jamaica by I. Drori during 1980–1 and 1984–5. The data were collected mainly through participant observation and household surveys.
2 A subclan; the exogamous social unit in Fiji which is the recognized primary land-owning unit.
3 Administratively, Fiji is made up of fifteen provinces divided among four divisions. Nandroga/Navosa is one of the three provinces which make up the Western Division.
4 The Agricultural Census figure of 1.75 acre may be an underestimated one due to villagers' tendency to record the area under cultivation as their total holding.
5 The requirement by the Fiji Development Bank for granting a loan is a security such as a lease, even on the farmer's own land. This lease is given by the Native Land Trust Board, but rarely for traditionally *matangali*-owned land.
6 The 1982 cash income data are assumed to be underestimated by up to 20 per cent. Despite such deficiencies, the data reasonably show the patterns of distribution of income sources and the degree of internal variation in cash income. Belshaw qualified his data similarly.
7 US $1 = J $1.78 at the time of the field-work (1980–1).

REFERENCES

Beckford, G.L. (ed.) (1975a) *Caribbean Economy: Dependence and Backwardness*, Mona: University of the West Indies, Institute of Social and Economic Research.
Beckford, G.L. (1975b) 'Caribbean rural economy', in Beckford (ed.) (1975a) *op. cit.*: 77–92.

Belshaw, C.S. (1964), *Under the Ivi Tree*, Berkeley and Los Angeles: University of California Press.

Benedict, B. (1967) *Problems of Smaller Territories*, London: Athlone Press.

Brookfield, H.C. (1972a) *Colonialism, Development and Independence: The Case of the Melanesian Islands in the South Pacific*, Cambridge: Cambridge University Press.

Brookfield, H.C. (1972b) 'Intensification and disintensification in Pacific agriculture', *Pacific Viewpoint* 14(1): 30–48.

Brookfield, H.C. (1979) 'Land reform, efficiency and rural income distribution: contributions to an argument', *Pacific Viewpoint* 20(1): 33–52.

Chandra, S. (1983) *Agricultural Development in Fiji*, Canberra: Australian Universities' International Development Program.

Connell, J. (1981) 'Diets and dependency. Food and colonialism in the South Pacific', Sydney: Freedom of Hunger Ideas Centre, Occasional Paper no. 1.

Connell, J. (1988) *Sovereignty and Survival: Island Microstates in the Third World*, Sydney: Research Monograph no. 3, Department of Geography, University of Sydney.

Demas, W. (1965) *The Economics of Development in Small Countries: with Special Reference to the Caribbean*, Montreal: McGill University Press.

Drori, I. (1982) 'The organization of production within agricultural cooperative in Jamaica', in H. Blustein and E. Le-Franc (eds) *Strategies for Organization in Small Farms in Jamaica*, Mona: Institute for Social and Economic Research, University of the West Indies and the Center of International Studies, Cornell University.

Drori, I. (1990) 'Land settlement in Jamaica: the implementation of socialist experience', *Public Administration and Development* 10: 27–39.

Drori, I. and Gayle, D.J. (1990) 'Youth employment strategies in Jamaican sugar-belt area', *Human Organization* 49(4): 364–72.

Dunham, D. (1982) 'Politics and land settlement schemes in Sri Lanka', *Development and Change* 13(1): 43–61.

Gayle, D.J. and Drori, I. (1984) 'Project OASIS: a case study in Jamaican development administration, *International Journal of Public Administration* 6(1): 97–125.

Hulme, D. (1984) 'Land settlement schemes and rural development in Papua New Guinea', unpublished PhD thesis, James Cook University, Townsville, Australia.

Hulme, D. (1987) 'State-sponsored land settlement policies: theory and practice' *Development and Change* 18(3): 413–36.

Hulme, D. (1988) 'Land settlement schemes and rural development: a review article', *Sociologia Ruralis* 28(1): 42–61.

Land Development Authority (LDA) (1963) *Annual Report for the Year 1962*, Suva: Legislative Council of Fiji, Council Paper no. 30.

Levitt, K. and Best, L. (1975) 'Character of Caribbean economy', in Beckford (ed.) (1975a), *op. cit.*: 34–61.

Ministry of Agriculture and Forestry (MAF), Fiji (1982) 'A Record of Attempts to Revive Defunct Land Development Authority Scheme Farms' (mimeo).

Manely, M. (1974) *The Politics of Change: A Jamaican Testament*, London: Andre Deutsch.

Maos, J.O. (1984) *The Spatial Organization of New Land Settlement in Latin America*, Boulder, Colorado: Westview Press.

National Planning Agency (NPA) (1978) *Five Year Development Plan 1978-82*, Kingston, Jamaica: Government Printer.

Oberai, A.S. (1986) 'Land settlement policies and population redistribution in developing countries, problems and prospects', *International Labour Review* 125(2): 141-61.

Rothfield, R. and Kumar, B. (1980) *Report on the Census of Agriculture, 1978*, Parliament of Fiji, Parliamentary Paper no. 28, Suva: Government Printer.

Schaaf, T. and Manshard, W. (1989) 'The growth of spontaneous agricultural colonization in the border area of Ghana and Ivory Coast', *Applied Geography and Development* 34: 7-22.

Sofer, M. (1987a) 'Uneven development in Fiji: a critical approach to the core-periphery concept', unpublished PhD thesis, University of Melbourne, Australia.

Sofer, M. (1987b) 'Progress through transformation – a Fijian village', *Pacific Viewpoint* 28(1): 1-19.

Sofer, M. (1988) 'Core-periphery structure in Fiji', *Environment and Planning D: Society and Space* 6: 55-74.

11

BREAKING OUT OF IMPORT-SUBSTITUTION INDUSTRIALIZATION

The case of Fiji

Rajesh Chandra

INTRODUCTION

Nearly all developing countries began their industrialization through import substitution (see Balassa 1981; Colman and Nixson 1978; World Bank 1987). While achieving rapid growth during the first phase of import substitution, lack of competition due to very high levels of protection produced inefficient industries, often relying on government subsidies for survival. Some countries, particularly those now known as Newly-Industrializing Countries (NICs), changed to export-oriented industrialization (EOI) in the 1960s, while many others did not do so until very much later. On the whole, countries adopting export-oriented industrialization have performed much better in output, productivity and exports than those which continued with import-substitution industrialization (ISI). In the 1980s, therefore, most developing countries have reviewed their industrial policies and, on the whole, changed towards EOI).

Fiji, a small, developing island state, less able to sustain a domestic market-oriented industrial strategy, began its industrial development after gaining independence in 1970 firmly within the ISI tradition. While the range of manufactured goods increased, there was little structural change in the economy, and in the 1980s in particular, the industrial sector stagnated (see Chandra 1985; Narsey 1989; World Bank

1986). The government had been making piecemeal changes to its industrial policies for some time, assisting export production with tariff rebates and offering companies tax benefits on export income. However, these changes were not effective; the export performance of the manufacturing sector continued to be dismal. In 1981 Fiji joined most of the small South Pacific developing countries in signing the South Pacific Regional Trade and Economic Co-operation Agreement (SPARTECA) between themselves and Australia and New Zealand. SPARTECA allowed duty- and quota-free entry of most island products to Australia and New Zealand. Fiji garment manufacturers were extremely quick to take advantage of these market opportunities. Later, in May 1987, a military coup deposed the government of the late Timothy Bavadra (Cole and Hughes 1988; Kasper *et al.* 1988; Roberts 1987; Robertson and Tamanisau 1988; and Rothwell 1987). In the wake of economic devastation, and fearing imminent economic collapse, the government devalued the currency by 33 per cent; introduced a tax-free system; and generally liberalized the economy. These changes reinforced those already underway, both because of SPARTECA and also as a result of Australian and New Zealand trade policies, and have turned the manufacturing sector from import-substitution to export-led growth. Exports of some manufactured goods, especially garments, have risen dramatically.

This chapter examines the transition in Fiji's industrialization, identifying the main features of the new policy. Following this introduction, the chapter briefly examines Fiji's manufacturing sector on the eve of independence, leading to a description of the increasing intensity of import substitution. The government had realized some of the problems its policies were creating, and had begun to make some changes in the early 1970s; this is examined in the next section. The chapter then argues that these changes were too feeble, and did not avert industrial stagnation. The role of SPARTECA and post-coup changes in industrial policies, especially the devaluation of the currency and the introduction of the tax-free system, that have dramatically changed the orientation and performance of the industrial sector are finally analyzed. Fiji's experience, although unique in some respects, can be used by other developing countries, especially small island countries.

MANUFACTURING SECTOR ON THE EVE OF INDEPENDENCE

Although the colonial government in Fiji, as elsewhere, did not favour industrialization, some industrialization took place because of the need to partially process commodities prior to export. In addition, during the war, manufacturing was encouraged to overcome shipping difficulties. After the war, the government continued its slight encouragement of industrialization, although without much determination or enthusiasm. With increases in standards of living also, additional manufacturing took place, mainly in low-value items, such as soft drinks.

The actual content and nature of the manufacturing sector on the eve of independence is indicated by an examination of the census of industries for 1968 (Fiji Bureau of Statistics 1970). The main features of the manufacturing sector in 1968 are summarized in Table 11.1. There were 310 establishments in the manufacturing sector in 1968, pointing to a major development of the manufacturing sector in the 1960s. Many of these establishments, however, were small and involved more in repair than in fabrication. The dominant component of the manufacturing sector was food processing, led by sugar and copra. Food-processing industries accounted for 80 per cent of the gross output. The other major industries related to sawmilling and its associated activities such as furniture and upholstery. On the whole then, the manufacturing sector had grown, but was largely confined to the processing of the primary sector output. Import-substituting industries had begun to develop but still remained nascent.

INTENSIFICATION OF IMPORT SUBSTITUTION

The colonial government itself began the process of ISI. In 1947 it had introduced the Protected Industries Ordinance, which empowered the governor to provide protection to industries he considered to be in the national interest. Although only three industries had been given assistance under this scheme by 1960, all of which had failed, it set the pattern for subsequent industrial development in Fiji; it was natural that newly-independent Fiji should follow this path. Moreover, although some countries, such as South Korea, had already

Table 11.1 The structure of the manufacturing sector of Fiji, 1968: number of establishments, employees and gross output[2]

Manufacturing subgroup	Establishments		Employees		Gross output (F$)	
	Number	Per cent	Number	Per cent	Amount	Per cent
Meat	25	8	270	3	941,297	2
Bakery products	17	5	325	3	1,337,523	2
Other foods, including prepared animal feeds	16	5	4,394	47	43,274,038	69
Drinks and tobacco	14	5	405	4	4,162,603	7
Tailoring	23	7	322	3	432,014	1
Footwear	11	4	31	<1	64,883	<1
Sawmilling, window and door frames, plywood, hardboard, and particle board	47	15	714	8	1,673,532	3
Curios	9	3	108	1	154,788	<1
Wooden furniture and upholstery	56	18	728	8	1,352,807	2
Paper and paper boards, printing, publishing and allied industries	21	7	489	5	1,294,959	2

Paints, varnishes, lacquers, soap and cleansing preparations	6	2	212	2	1,729,262	3
Rubber and plastic products	6	2	93	1	170,158	<1
Cement; ready-mixed and concrete blocks	11	4	285	3	1,986,244	3
Furniture and fixtures of metal, electroplating, structural metal products	20	6	358	4	1,263,941	2
Agricultural and industrial machinery, including repairs and renovations	18	6	414	4	1,786,346	3
Ships and boats	8	3	156	2	548,331	1
Items not elsewhere classified	2	1	102	1	127,814	<1
TOTAL	310	101	9,406	99	62,300,540	100

Source: Calculated from data in Fiji Bureau of Statistics 1970: 36.
Note: a These figures have been grossed-up to take account of non-response to the Bureau census.

switched their industrial strategies towards export production by the mid-1960s, development thinking still did not favour export orientation universally. In any case, Korean industrial success had not become widely known.

The government followed an incremental policy for the manufacturing sector in the 1970s and 1980s. The manufacturing sector was not seen as a leading sector of the economy; this role was accorded to agriculture and tourism. The government did, however, recognize its importance, and formulated policies to encourage it.

The government viewed manufacturing as an area for the private sector, with only indirect government action (Fiji Central Planning Office 1970). The state would upgrade manpower training, create favourable investment climate, offer concessions, develop industrial estates, and provide a range of extension services, much like its agricultural extension services.

The government consolidated its industrial polices with the introduction of the new Income Tax Act in 1974. The Act set out assistance to the commercial sector in schedules 3 and 5. Under the new Act, where enterprises were deemed to assist the economic development of Fiji, exemptions from tax were granted (Fiji Government 1975). Exemptions were to be for five years initially, with provision for extension (ibid.).

The tax exemption related to profits or gains of F$5,000 per annum or, if greater 15 per cent of the smaller of the paid-up equity capital of the company or of the total fixed capital investment of the company (Fiji Government 1975). Furthermore, losses incurred by the approved enterprise could be carried forward for six years after the expiry of the tax exemption period.

In addition, the government continued to levy high rates of tariffs on imports, particularly where these items were being produced in Fiji. During this period (up to the late 1980s) the government looked upon tariffs as essential revenue generators. In 1983, for instance, customs and port dues accounted for 35 per cent of its current revenue. Additionally the government imposed licenses and quotas on a wide range of products, often in response to direct pressure from manufacturers. By the early 1980s, the Fiji Employment and Development Mission (FEDM) pointed out that protection had reached 'dangerously high levels') (FEDM 1984:510). The

The World Bank (1986) also pointed to very high levels of effective protection, even leading to negative value added in some instances (Narsey 1989).

FEEBLE ATTEMPTS AT REFORM

There are indications that the government had realized the importance of reforming its industrial policy at an early stage, and wished to see exports from the initially import-substituting industries. Even Development Plan Six (1971–6), the first for independent Fiji, argued that because of international competition and its small domestic market, Fiji should concentrate on a narrow range of manufacturing and processing activities which had moderate-scale economies, some export potential, especially in the Pacific Basin, and which could enhance value-added to presently-exploited and potentially-exportable domestic resources (Fiji Central Planning Office 1970). The fact, the Plan marked Fiji's first major attempt at shifting the orientation of manufacturing in Fiji through its proposal to provide assistance to and promote export-oriented manufacturing. It also abolished, in 1970, an existing export tax of 2 per cent except for sugar and coconut oil, the country's principal exports. The government also said that it intended to promote import-substituting industrial development in areas where export potential was present and it indicated its wish for more manufactured exports:

> As industrial development progresses and output of manufactured goods increases, Fiji will concentrate on securing an increasing outlet for her manufactures in the more competitive markets of the developed countries as well as in the markets of other less developed countries.
>
> (Fiji Central Planning Office 1970: 70)

The change in attitude in Fiji toward export production is interesting as there has been a marked international change, from the mid-1960s, towards an export orientation in manufacturing in a large number of developing countries (see Donges 1983). In Taiwan and Korea, for instance, a marked export orientation was signalled in the 1958–61 period, when the incentive structure was overhauled (Balassa 1981:61; see also Ho 1978 and Kuo 1981).

211

As part of the new Income Tax Act, the government also provided export incentives. All export products, except sugar and copra, were eligible, but the total export sales had to be more than F $5,000 (US $2,000). A complex graded system of rebate on taxes on export earnings was introduced, which could be granted for seven years. However, the incentives were insufficient given that extremely high levels of protection made the local market a more profitable and easy option. Some firms were attempting to export, but their main market remained Fiji and the export performance was erratic and, almost without exception, deteriorated over the 1970 to 1986 period. Furthermore, the overvalued Fijian dollar, attributable to the import-substitution policies of the government, meant that whatever export potential Fijian manufacturers had was not fully utilized.

In Development Plan Eight (1981–5), the government did promise to set up a trade and marketing authority, to negotiate bilateral marketing agreements, and to create an export-processing zone. The Economic Development Board was created to assist with investment promotion and international marketing. But the government, which had been considering the issue of an export-processing zone since 1973, did not make any decisions on the issue.

In Development Plan Nine (1986–90), the government once again showed its awareness of the need for change in the industrial sector. For instance, it identified the creation of a dynamic, growth-oriented sector as one of the five main objectives of the sector; it also wished to expand output for export; encourage private investment; and to increase the efficiency and effectiveness of the manufacturing sector. These objectives were to be achieved through improved government services, including the provision of industrial land, and improved bureaucracy, new institutions, and a commitment to a more open economic system. However, as has already been indicated, the Plan was aborted in early 1987.

STAGNANT INDUSTRIAL SYSTEM

There is widespread agreement that until the late 1980s, the manufacturing sector in Fiji was stagnant (see Chandra 1985; FEDM 1984; Narsey 1989; World Bank 1986). We can see this stagnation in the failure of manufacturing to increase its

Table 11.2 Gross domestic product at factor cost by economic sector, 1950–70 at current prices (F$ million) with percentage share in parenthesis

Sector	Period					
	1950	1953	1957	1963	1965	1970
Primary sector						
Agriculture, forestry, and fishing	15.9 (44)	26.0 (51)	29.2 (48)	44.6 (41)	40.6 (34)	54.7 (29)
Mining and quarrying	1.9 (5)	1.5 (3)	1.9 (3)	1.8 (2)	2.2 (2)	2.4 (1)
Secondary sector						
Manufacturing & processing	3.9 (11)	4.6 (9)	6.3 (10)	13.3 (12)	14.9 (12)	23 (12)
Construction, electricity, and water	2.0 (6)	2.7 (5)	4.3 (7)	6.8 (6)	8.0 (7)	16.7 (9)
Transport and communications	1.0 (3)	1.9 (4)	2.4 (4)	7.8 (7)	6.8 (6)	14.8 (8)
Tertiary sector						
Government services	3.3 (9)	3.4 (7)	3.5 (6)	5.9 (5)	8.5 (1)	13.3 (7)
Other services[a]	8.0 (22)	10.7 (21)	13.5 (22)	28.3 (26)	40.1 (33)	63.5 (34)
GDP	36.0 (100)	50.8 (100)	61.1 (100)	108.5 (99)	121.1 (101)	188.5 (100)

Sources: Fiji Central Planning Office 1970: 4; 1980: 1 (for 1970 data). Percentages have been calculated by the author.

Note: Includes commerce, tourism, real estate, and property ownership.

contribution to the gross domestic product (GDP); in almost static manufacturing employment and value added; and in erratic and insignificant exports from the manufacturing sector.

Tables 11.2 and 11.3 present data on structural change in the Fiji economy. It is clear that the manufacturing sector not only did not increase its share of the GDP, but it actually fell in some years. Indeed, it is worth noting that the manufacturing sector's contribution to GDP was lower in 1988 than in 1950. Data on manufacturing employment are presented

Table 11.3 Fiji: share of the manufacturing sector of the GDP at current prices at factor cost, 1980–8

Year	Share of the GDP
1980	11.9
1981	10.5
1982	10.7
1983	9.2
1984	9.8
1985	9.5
1986	10.2
1987	10.2
1988	9.9

Sources: Calculated from Fiji Bureau of Statistics 1989a: 7 and Fiji Bureau of Statistics 1991: 7.

Table 11.4 Fiji: real value added per capita, 1973–85 (F$ '000)

Year	Value added	Percentage change
1973	3.23	---
1974	3.96	22.6
1975	4.04	2.0
1976	380	– 5.9
1977	4.03	6.1
1978	3.47	– 13.9
1979	4.28	23.3
1980	3.68	– 14.0
1981	3.37	– 8.4
1982	3.32	– 1.5
1983	2.54	– 23.5
1984	2.75	8.3
1985	2.70	– 1.8

Source: Narsey 1989:39. Percentages have been calculated by the author.

in Figure 11.1, from which it is clear that manufacturing employment only increased marginally between 1974 and 1988 and for several years in the 1980s; the absolute size of manufacturing sector employment actually declined. Figure 11.1 and Table 11.4 present data on the growth of manufacturing value added, and indicate yet again industrial stagnation. Perhaps the greatest indictment of the manufacturing sector is to be found in its dismal export performance, presented in Table 11.5. Firms that were given heavy protection simply did not export; whatever exports took place went to the island South Pacific, and performance was extremely erratic.

Table 11.5 Fiji: export performance of import-substituting manufacturing, 1970-90

Year	Cement (000 tonnes)	Bakery products (000 kg)	Cigarettes (000 kg)	Paints (000 litres)
1970	11	7i6	67	164
1971	24	913	67	168
1972	14	893	54	186
1973	12	786	28	241
1974	15	1,074	11	177
1975	12	724	5	123
1976	10	752	8	116
1977	8	671	17	151
1978	10	744	9	154
1979	3	940	6	137
1980	2	827	6	124
1981	1	848	5	109
1982	0	787	3	193
1983	2	1,003	9	229
1984	4	1,013	3	162
1985	8	857	5	217
1986	3	901	6	208
1987	3	676	4	212
1988	4	678	10	219
1989	21	1,256	12	138
1990	30	1,105	10	200

Sources: Fiji Bureau of Statistics 1981:56-7; 1984:59-60; 1989a:60-1; 1991:60-1.

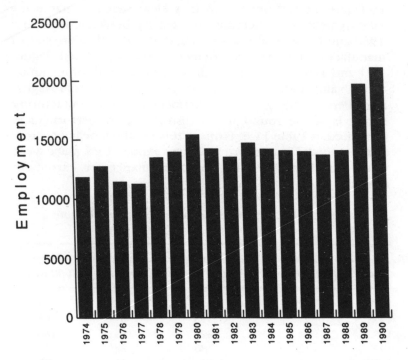

Figure 11.1 Fiji: manufacturing sector employment, 1974–90

SOUTH PACIFIC REGIONAL TRADE AND ECONOMIC CO-OPERATION AGREEMENT

One of the most crucial events affecting the future of manufacturing in the South Pacific was the implementation of SPARTECA. It became effective in January 1981, but there were initial restrictions on some manufactured goods, such as textiles, clothing, and footwear (see Thomson 1989). However, New Zealand decided to allow completely duty-free entry for all island goods, while Australia decided to provide duty-free entry for a fixed quota.

Industrial stagnation explained

The overall reason for industrial stagnation in Fiji was clearly the government's adherence to ISI. Once the easy phase of

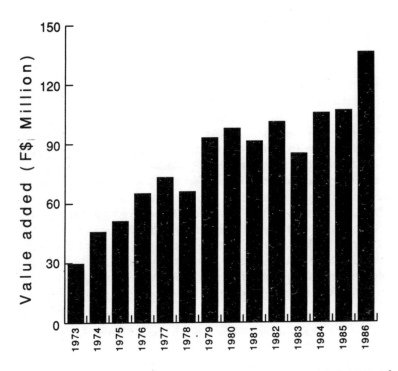

Figure 11.2 Fiji: growth of manufacturing sector value added, 1973–86

ISI was over, the government should have reoriented manufacturing towards exports. But this did not happen. Hence output could not increase because the domestic market was saturated; inefficient industries meant that exports were not viable; and the small market meant that the next round of ISI was prohibitive.

But there were other reasons for industrial stagnation. The government did not have a definite long-term industrial policy; interviews with over one hundred manufacturers in 1983 indicated that they considered this to be an important shortcoming on the part of the government (Chandra 1985). The government also showed ambivalence towards foreign investment. For instance, it openly courted select local firms, and severely limited access to the cheap local capital markets. According to Taylor (1983:51), the government also did not provide any assistance to foreign corporations in dealing with the problems of mushrooming small local firms directly competing with them, and paying wages sometimes as low as one-fifth of those paid by foreign corporations. There was, thus, an unattractive investment climate in Fiji in the late 1970s and early 1980s (see also Bowden 1976). This led, not only to decreased foreign investment, but large-scale disinvestment by long-established companies.

This market access, combined with the gradual reduction of protection in both Australian and New Zealand markets meant that Australian and New Zealand manufacturers had to restructure their operations and seek arrangements for offshore production, given the high labour costs of these countries. In 1986, for instance, the Fijian manufacturing wages were only 15 per cent of those of Australia and 19 per cent of New Zealand (Chandra 1991:76). It is useful to note here that since New Zealand's decisions to open its market to Fiji garments, more than 20 per cent of its garment production has relocated to Fiji (*TCF Industry Advisor* 1989:4).

Table 11.6 Fiji: garment exports, 1975–90

Year	Value of exports (F$)	Per cent change
1975	58,464	---
1976	62,296	7
1977	70,889	14
1978	87,297	23
1979	119,557	37
1980	115,043	– 4
1981	108,751	– 5
1982	170,793	57
1983	516,675	203
1984	848,513	64
1985	2,058,809	143
1986	4,862,771	136
1987[a]	8,803,744	81
1988	18,222,129	107
1989[b]	97,305,792	434
1990	113,749,925	17

Sources: Fiji Bureau of Statistics 1986:191, and unpublished trade data from Fiji Bureau of Statistics 1991.
Notes: a The Fiji dollar was devalued by 33 per cent in 1987.
    b The government has revised garment export figures due to irregularities in customs entries. The new figures are more accurate.

Some indication of the importance of SPARTECA can be gained by examining the growth of garment exports from Fiji, given in Table 11.6.

It needs to be emphasized that the first major spurt of growth in garment exports occurred in 1982; SPARTECA had already become effective in 1981. We can also note the very significant growth in exports from 1985 onwards. Clearly, garment exports were set to grow explosively in the aftermath of the opening of the markets by New Zealand and Australia.

POST-COUP CHANGES

Devaluation of the currency

In an attempt to deal with an unprecedented pressure on foreign reserves and to revive the tourist industry, the government decided to devalue the currency by 33 per cent in 1987. It can be argued that this sudden reduction in the foreign exchange costs of Fiji products made exporting much easier. It undoubtedly did so, but the fact that other factors are important is indicated by the fact that few other areas have succeeded in exports apart from garments. The real significance of the devaluation lay in the government's successful effort in bringing about a decline in real wages. This has effectively depressed wages to those of a few years ago (Keith-Reid 1989).

It is interesting to note that after significant increases in the 1970s, particularly in early 1970s, wage increases had declined in the 1980s (Table 11.7). Moreover, even absolute wages had begun to decline. For instance, the mean daily wage in the manufacturing sector in 1986 was lower that in 1983. In real terms, manufacturing wages had begun to fall from 1978.

Tax-free system

After vacillating on the issue for almost a decade and a half, the tax-free system was introduced in late 1987. For manufacturers and export-service providers the government offered an attractive package, provided they exported at least 95 per cent of their output (Chandra 1990b for a full description of the tax-free system).

Table 11.7 Daily mean manufacturing wages in Fiji, 1965–88

Year (end of June)	Wage (F$)	Percentage increase over previous year	Annual rate of inflation
1965	1.98	---	na
1966	2.13	7.6	na
1967	2.20	3.3	na
1968	2.28	3.6	na
1969	2.44	7.0	3.9
1970	2.55	4.5	4.1
1971	2.69	5.5	6.5
1972	3.21	19.3	9.1
1973	3.78	17.8	11.2
1974	4.82	27.5	14.4
1975	6.19	28.4	13.1
1976	6.64	7.3	11.4
1977	6.87	3.5	7.0
1978	8.10	17.9	6.1
1979	8.48	4.7	7.7
1980	9.52	12.3	14.5
1981	10.56	10.9	11.2
1982	11.20	6.1	7.0
1983	11.92	6.4	na
1984	12.00	0.7	na
1985	12.16	1.3	4.4
1986	11.84	-2.6	1.8
1987	12.32	4.1	5.7
1988	12.48	1.3	11.9

Sources: Fiji Bureau of Statistics, 1984:66,78; 1989a:78; 1991:69,79.

The investment and incentives package involves a total waiver of licensing for import of capital goods and other production materials; duty-free import of capital goods and equipment; customs duty exemption on importation of raw materials, components, spares and packaging materials. Building materials, furniture and office equipment used to establish a tax-free factory are also exempt from customs duty. The government also exempts from excise duty products manufactured within the zone and treats all materials supplied to the zone from outside it, as exports, and therefore eligible for export benefits such as remission of excise duties (Fiji Trade and Investments Board 1988:31).

Companies are offered a tax holiday of thirteen years; there is no withholding tax on interest, dividends and royalties paid abroad; final dividend is taxed at the rate of 5 per cent when paid to resident shareholders; and there are no restrictions on repatriation of capital and profits.

The tax-free system also offered further streamlining of the bureaucratic procedures, with the Fiji Trade and Investment Board (FTIB) handling most of the paperwork on behalf of the investor. The FTIB has promised a turnaround time of six weeks from application to notification, though in practice a turnaround time of about ten weeks now prevails.

In addition, the tax-free status entitles firms to import specialist personnel for enterprises subject to requirements under the Immigration Act. The immigration rules have been considerably relaxed to encourage foreign investment.

Deregulation

The government has deregulated the economy through a major reduction in the list of items covered. In July 1989, import licence controls were lifted on thirty-four non-agricultural products; of the remaining fifteen mainly agro-based products remaining under licence control in 1989, seven were removed in the 1991 budget (Fiji Ministry of Finance and Economic Planning 1991:70). Additionally, the government has committed itself to a reduction in the fiscal duties on imported goods also produced domestically. In the 1991 budget, fiscal duties on most such goods were reduced to 40 per cent and further reductions of 10 per cent are scheduled for 1992 and 1993.

Other measures

In addition to the above, the government has also embarked on a taxation reform, which has already reduced the highest personal tax rate by 20 per cent and which is expected to see the introduction of a value-added tax in July 1992. The government has also given notice of its intention to reform the labour market in line with its policy of building a competitive, external-oriented economy.

The government has also declared that technology and productivity will be its themes in the 1930s (Kamikamica 1989:19). In support of its strategy, the government has reduced duties on computers and telecommunications equipment.

RESULTS OF RECENT CHANGES IN THE MANUFACTURING SECTOR

We have already seen the dramatic rise in garment exports from Fiji, reaching F $114 million in 1990 (Table 11.6). The Secretary for Commerce and Industry, Navitilai Naisoro, projects that garment exports will reach F $300 million by 1993/1994 and F $1 billion by A.D. 2000 (Keith-Reid 1990:3–4). Such is the optimism in the garment sector that Maurice Lubansky, head of one of the largest garment manufacturers, says that as South Korea, the world's garment hub for the last quarter century, has become uncompetitive, 'Fiji can be the replacement for South Korea for the next 25 years' (quoted in Keith-Reid 1990:4).

Although garment exports have been the most spectacular, exports of other products in the tax-free sector have also been increased. Even if we discount the greatly-increased exports of the Pacific Fishing Company because the granting of tax-free factories (TFF) status is not likely to have assisted in this expansion, exports of cement, lumber, cigarettes, paints and veneer have all shown promising increases. New exports of leather goods, knock-down furniture, and buttons and blankets have also developed.

Foreign investment, which had come to almost standstill in the 1980s, has increased significantly, and is likely to increase even more in the future provided political stability continues. The tax-free sector is clearly set for a major expansion in the immediate future. At the end of March 1990, ninety-eight tax-free factories, mostly in garment production, were operational, employing 8,226 people, and representing an investment of F $36.8 million. Some indication of the importance of the tax-free factory in Fijian manufacturing is given by the fact that its labour force represented almost 40 per cent of the total manufacturing labour force (for more detail of the tax-free system, see Chandra 1990b).

CONCLUSIONS

Fiji, which retained an ISI strategy into the 1980s, suffered most of the problems associated with it. Given the small size of the domestic market, ISI created more crippling problems than in larger countries.

The government, which offered strong inducements to manufacturers for domestic market production, failed to show any boldness in dealing with problems. The planning office showed an early recognition of the problems of ISI; this is well documented in the development plans. The government also showed that it was aware of the importance of fostering competition, and of exporting. So why did it not change its policies?

The government did try to improve the industrial system. For instance, it created new institutions, such as the Economic Development Board; it tried to encourage exports by abolishing export tax; and it offered some additional fiscal incentives for exporters. However, the government failed to make a decisive break with the ISI; minor tinkering did nothing to change the fact that manufacturers catering to the domestic market were extremely well off; there was no reason for them to take risks associated with exporting. At least some of the government's poor performance can be attributed to the fact that being backed by manufacturers, it could not see its way clear to effectively reduce their profits; there was a very powerful and effective business lobby.

The first major impetus to exporting came with the coming into force of SPARTECA. We cannot overemphasize the importance of this concessionary market arrangement. Fiji's garment manufacturers, who had enjoyed very significant government assistance, had long argued that they could export; with SPARTECA in 1981, they got the opportunity. Thus SPARTECA is the real reason for the dramatic growth of garment exports to Australia and New Zealand.

Allied with SPARTECA is the intention of both Australia and New Zealand to reduce levels of protection for their industries. This has forced Australian and New Zealand manufacturers to look for offshore production and other marketing and production arrangements. With this interest already present in these countries, Fiji's tax-free system, and its very effective marketing in these countries, Fiji was set to receive significant foreign investment, which has also led to further increase in exports. Thus, while exports were increasing, they have increased faster with the coming of Australian and New Zealand (and other) investors. The government's devaluation of the currency has also helped in the short term in making Fiji's garments relatively cheaper.

It should be noted that Fiji's future industrial and economic performance depends on her finding a broadly acceptable constitutional accommodation among the main ethnic groups; stability is crucial to ensure long-term investments. Moreover, the government has to capitalize on the opportunities created by recent industrial expansion to seek a diversified range of industries catering to a wider market than those currently served (Australia and New Zealand).

Although there are many peculiarities regarding Fiji's recent transition from import-substitution to export-oriented manufacturing, such a relatively well-developed industrial base relative to other South Pacific countries and the dramatic impact of the military coups, some lessons could still be learnt by Fiji's neighbours. First, import-substitution industrialization strategy entails considerable costs and is not viable for small countries. Second, if a shift from import-substitution is contemplated, then bold and decisive policies will be needed, such as the inertia of ISI. Third, other South Pacific countries need to introduce a better type of tax-free system if they are to be able to compete for scarce foreign investment. Fiji has marketed its incentives very effectively, and other countries can learn from its Trade and Investment Board. Finally, Fiji has demonstrated that a small island country can succeed with the right mix of external opportunity, public policies, and domestic entrepreneurs. While other South Pacific countries do not have all the favourable conditions Fiji had, they can derive more hope for the industrialization of their societies.

REFERENCES

Balassa, B. (1981) *The Newly Industrializing Countries in the World Economy*, Oxford: Pergamon Press.

Bowden, E.G. (1976) *Manufacturing Employment Potential in Fiji: Analysis and Policy Recommendations*, Discussion Paper no. 4, Suva: United Nations Fiji Regional Planning Project.

Carstairs, R.T. and Prasad, R.D. (1981) *Impact of Foreign Direct Investment on the Fiji Economy*, Suva: Centre for Applied Studies in Development, University of the South Pacific.

Chandra, R. (1985) 'Industrialization in Fiji', unpublished PhD thesis, Vancouver: Department of Geography, University of British Columbia.

Chandra, R. (1990b) *Fiji's Tax Free System*, Honolulu: Pacific Islands Development Program, East-West Center.

Chandra, R. (1991) 'Industrialization in the Pacific islands: an overview', forthcoming occasional paper, Pacific Islands Development Program, East-West Center.

Cole, R. and Hughes, H. (1988) *The Fiji Economy, May 1987. Problems and Prospects*, Pacific Policy Papers no. 4, Canberra: National Centre for Development Studies, The Australian National University.

Colman, D. and Nixson, F. (1978) *Economies of Change in Less Developed Countries*, Oxford: Philip Allan.

Dicken, P. (1986a) *Global Shift: Industrial Change in a Turbulent World*, Cambridge: Harper & Row.

Donges, J.B. (1983) 'Re-appraisal of foreign trade strategies for industrial development, in G.F. Machluf and H. Muller-Groeling (eds) *Reflections on a Troubled World Economy. Essays in Honour of Herbert Giersch* London and Basingstoke: Macmillan for the Trade Policy Research Centre, 279–301.

ESCAP/UNCTC (1985) *Transnational Corporations and the Developing Pacific Island Countries*, Bangkok: United Nations.

Fiji Bureau of Statistics (1970) *Census of Industries, 1968*, Suva: Fiji Bureau of Statistics.

Fiji Bureau of Statistics (1981) *Current Economic Statistics, January 1981*, Suva: Fiji Bureau of Statistics.

Fiji Bureau of Statistics (1984) *Current Economic Statistics, July 1984*, Suva: Fiji Bureau of Statistics.

Fiji Bureau of Statistics (1986) *Overseas Trade, 1985*, Parliamentary Paper no. 49 of 1986, Suva: Government Printer.

Fiji Bureau of Statistics (1988) *Census of Industries, 1985*, Suva: Fiji Bureau of Statistics.

Fiji Bureau of Statistics (1989a) *Current Economic Statistics, July 1989*, Suva: Fiji Bureau of Statistics.

Fiji Bureau of Statistics (1989b) *Social Indicators of Fiji*, no. 5, Suva: Fiji Bureau of Statistics.

Fiji Bureau of Statistics (1991) *Current Economic Statistics, April 1991*, Suva: Fiji Bureau of Statistics.

Fiji Central Planing Office (1970) *Fiji's Sixth Development Plan, 1971–5*, Parliamentary Paper no. 25 of 1970, Suva: Government Printer.

Fiji Central Planning Office (1979) *A Review of Fiji's Seventh Development Plan, 1976–8*, Suva: Fiji Central Planning Office.

Fiji Central Planning Office (1980) *Fiji's Eighth Development Plan, 1981–5*, Vol. 1, *Policies and Programmes for Social and Economic Development*, Suva: Fiji Central Planning Office.

Fiji Central Planning Office (1981) *Development Plan Eight Annual Review*, mimeo, Suva: Fiji Central Planning Office.

Fiji Central Planning Office (1985) *Fiji's Ninth Development Plan, 1986–90*, Suva: Fiji Central Planning Office.

Fiji Employment and Development Mission (FEDM) (1984) *Final Report to the Government of Fiji*, Parliamentary Paper no. 66 of 1984, Suva: Government Printer.

Fiji Government (1975) *Laws of Fiji 1974*, Suva: Government Printer.

Fiji Ministry of Finance and Economic Planning (1991) *Review of Performance and Prospects: Fiji Economy*, Suva: Fiji Ministry of Finance and Economic Planning.

Fiji Trade and Investment Board (1988) *Fiji Investment Guide*, Suva: Fiji Trade and Investment Board.

Garrison, R. (1990) 'The garment industry in the South Pacific', *Tok Blong SPFF* 30: 26-9.

Gunasekera, H.M. (1985) 'Fiji', in ESCAP, *Patterns and Impact of Foreign Investment in the ESCAP Region*, Bangkok: ESCAP, 27-41.

Ho, S.P.S. (1978) *Economic Development of Taiwan, 1860-1970*, New Haven, Connecticut: Yale University Press.

Kamikamica, J.N. (1988) '1988 mini budget statement by the Minister of Finance and Economic Planning, Hon. Josefata N. Kamikamica', Tuesday 28 June, mimeographed copy in the possession of the author.

Kamikamica, J.N. (1989) *Budget 1990*, Suva: Fiji Ministry of Finance and Economic Planning.

Kasper, W., Bennett, J. and Blandy, R. (1988) *Fiji: Opportunity from Adversity?* St. Leonards (NSW): The Centre for Independent Studies.

Keith-Reid, R. (1989) 'The new Fijian takes charge', *Islands Business* 15(2): 8-17.

Keith-Reid, R. (1990) 'The classic way for Fiji's garment industry to go', *Islands Business* February: 1-4.

Kreye, O., Heinrichs, J., and Frobel, F. (1987) *Export Processing Zones in Developing Countries: Results of a New Survey*, Geneva: International Labour Office, Working Paper no. 43, Multinational Enterprises Programme.

Kuo, S.W.Y. (1981) *Taiwan Success Story, Rapid Growth and Improved Distribution in the Republic of China, 1952-79*, Boulder, Colorado: Westview Press.

Nair, R. *et al.* (1986) *Report and Recommendations of Tribunal, Garment Manufacturing Industry*, Parliamentary Paper no. 38 of 1986, Suva: Government Printer.

Narsey, W.L. (1988) 'Labour absorption, incomes and productivities: differentials between foreign and local companies in Fiji manufacturing, 1973-85', Paper presented to Conference on Industrialization and Economic Development of South Pacific Island States: Problems and Prospects, University of the South Pacific, Suva, 12-15 December.

Narsey, W.L. (1989) *The Performance of the Fiji Manufacturing Sector, 1973-85*, Working Paper no. 14, Suva: School of Social and Economic Development, University of the South Pacific.

Ogilvie, D. (1989) 'Fiji's tax free zone/tax free factory package – an assessment of the first year's result', unpublished report to the Fiji Trade and Investment Board, Suva.

Prakash, J. (1982) 'Underdevelopment and industrialization in Fiji', Unpublished MA thesis, University of Dar-es-Salaam, Dar-es-Salaam, Tanzania.

Prasad, S. (1988) 'Tax free zones: systematic industrialization or programmed underdevelopment?', Unpublished review article in the possession of the author.

Qionibaravi, M. (1986) *Budget 1987*, Suva: Government Printer.

Roberts, M. (1987) 'Economies of the Pacific', *Pacific Islands Monthly* 58(11): 33–41.

Robertson, R.T. and Tamanisau, A. (1988) *Fiji: Shattered Coups*, Leichhardt (NSW): Pluto Press.

Rothwell, N. (1987) 'Rabuka's republic', *Pacific Islands Monthly* 58(11): 7–15.

Sklair, L. (1986) 'Free zones, development and the new international division of labour', *Journal of Development Studies* 22(4): 752–9.

Slatter, C. (1987) 'Women factory workers in Fiji: the "half a loaf" syndrome', *Journal of Pacific Studies* 13: 47–59.

Taylor, M. (1983) 'Foreign capital in the Fiji economy: context and concerns', Working Paper, Suva: Fiji Employment and Development Mission to the Fiji Government.

TCF Industry Advisor, 1(9) (1989).

Thomson, P.W. (1989) *Trade and Investment in the South Pacific Islands: A Diagnostic Study*, Honolulu: Pacific Islands Development Program, East-West Centre.

United Nations Industrial Development Organization (UNIDO) (1980) *Export Processing Zones in Developing Countries*, Working Papers on Structural Changes no. 19, Vienna: UNIDO.

Warr, P.G. (1987a) 'Export promotion via industrial enclaves: the Philippines' Bataan Export Processing Zone, *Journal of Development Studies* 23(2): 220–41.

Warr, P.G. (1987b) 'Malaysia's industrial enclaves: benefits and costs', in T.G. McGee, *et al. Industrialization and Labour Force Processes. A Case Study of Peninsula Malaysia*, Research Papers on Development of East Java and West Malaysia, no. 1, Canberra: Research School of Pacific Studies, The Australian National University, 179–215.

World Bank (1986) *Fiji: A Transition to Manufacturing*, Report no. 6372-FIJ, Washington DC: World Bank.

World Bank (1987) *World Development Report, 1987*, New York: Oxford University Press for the World Bank.

12

TOURISM AND POLITICS
The example of Cyprus
Douglas G. Lockhart

INTRODUCTION

Although a large volume of research has been published on
the characteristics of tourism on islands, much of this relates
to the structure of the industry, social, economic and
environmental impacts, and land use and visitor patterns
(Pearce 1987). The literature on political aspects is by comparison fairly slight (Butler 1993; Chapter 5). Both Bastin
(1984) and English (1986) have commented upon the implications for the industry caused by political instability in the early
1970s and the way dependency on foreign companies and
markets exacerbated existing political sensitivities. Surprisingly, military *coups d'état*, warfare and ethnic strife, while
having had the most dramatic impact upon tourist arrivals,
have received the least attention from academic researchers.
Lea (1988) has dealt with the downturn of tourism in Fiji
following the coups in 1987. The negative impact on Malta's
tourist industry of the American air force bombing raid on
Libya in 1986 attracted considerable attention from Maltese
economic planners but, as Lockhart and Ashton (1991) have
pointed out, this was just one of many factors influencing
tourist arrivals at the time. Other countries which once had
flourishing holiday industries such as the Lebanon and
Yugoslavia have seen these virtually disappear as a result of
prolonged hostilities yet have received scant commentary in
the geographical literature.

Cyprus, however, is unique, because it has witnessed
periodic dislocation of its tourist industry for more than
thirty years. Civil unrest during the 1950s and 1960s was
accompanied by downturns in arrivals and it has had to

reconstruct much of its tourist infrastructure following the *coup d'état* against the government in July 1974 which was followed by Turkish military intervention and the *de facto* division of the island into a Greek Cypriot controlled south (Republic of Cyprus) and a Turkish Cypriot administered north, known between 1975 and 1983 as the Turkish Cypriot Federated State, and since a unilateral declaration of independence in November 1983 it has been called the Turkish Republic of Northern Cyprus. Tourism in the two areas has developed along very different paths. Moreover, because of international travel restrictions, tourists wishing to stay overnight can only enter North Cyprus via Turkey and Syria. No such restrictions apply to the South which has direct access to Europe and the Middle East via airports at Larnaca and Paphos that are in close proximity to the major resorts.

EVOLUTION OF TOURISM, 1950–73

Although the post-1974 period is dominated by the repercussions of the Turkish military presence, tourism in earlier years was also affected by various internal political problems. At various times, a state of emergency (1956–8), the civil war in 1963 and 1964, and further fighting in 1967, had a marked impact upon business confidence and tourist arrivals. As a result the pattern of tourist arrivals was characterized by periods of expansion punctuated by recession, and it was only during the late 1960s and the early 1970s that rapid growth was achieved.

When independence from British rule was granted in 1960 arrivals amounted to only 25,700 and foreign exchange earnings were a mere CY £3 million. In common with other Mediterranean resorts, tourist arrivals increased steadily. (Figure 12.1). The intercommunal violence in December 1963 that led to the partitioning of the capital, Nicosia, and the creation of enclaves populated by the Turkish Cypriot minority was reflected in lower numbers of tourists during 1964. Friction between the two communities continued for several years. However, this was largely confined to remote areas rarely visited by tourists, although the large North Nicosia-Geunyeli (Gonyeli) enclave closed the direct route to the resort town

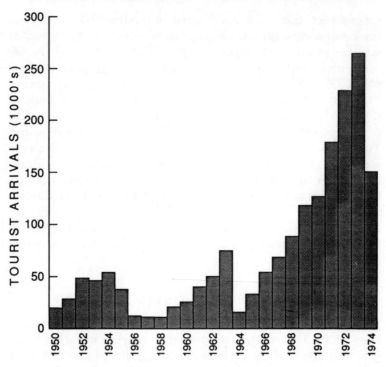

Figure 12.1 Tourist arrivals staying for more than one night, 1950-74

of Kyrenia, an obstruction which led to the construction of a new road through the Kyrenia mountain range near the Pendadhaktyos peak. Apart from warfare at Kokkina (Erenkoy) in the Tillyria mountains during August 1964 and a serious attack upon two Turkish-Cypriot villages near the Nicosia-Limassol road in 1967, tension between the two communities gradually eased and as a result tourism enjoyed boom conditions in the early 1970s. During the period which coincided with the Second Five Year Economic Plan (1966-71) the annual growth of world tourism was in the range of 3-10 per cent. In comparison the number of tourists arriving in Cyprus increased by 41 per cent in 1970 and 1971 which was one of the highest growth rates in the world. Earning from tourism increased substantially ahead of projected targets as a result of higher average length of stay, higher expenditure per

Table 12.1 Republic of Cyprus: tourist arrivals staying in all accommodation types by nationality, 1970–90 (%)

	UK	Greece	West Germany	Scandinavia[1]	Lebanon[2]	Other sources	Total arrivals
1970	47.4	7.4	2.3	2.3	7.1	33.5	126,580
1971	43.7	6.9	5.7	3.0	6.0	34.7	178,598
1972	43.0	5.6	8.5	6.8	4.6	31.4	228,309
1973	43.9	5.7	10.1	8.2	3.8	28.2	264,066
1974	35.2	6.4	11.2	11.8	5.0	30.4	150,478
1975	37.1	15.9	2.4	1.3	9.2	34.1	47,084
1976	19.1	8.7	2.4	1.1	43.2	25.4	180,206
1977	31.1	12.6	4.3	2.5	14.8	34.7	178,185
1978	34.4	11.8	4.9	4.8	11.9	32.2	216,679
1979	35.8	11.3	6.7	6.4	8.9	31.0	297,013
1980	31.5	9.5	8.1	10.7	6.4	33.8	353,375
1981	30.1	6.5	7.2	17.8	7.3	31.1	429,313
1982	24.6	6.3	6.5	19.9	9.6	33.2	548,180
1983	27.1	5.2	6.0	13.9	12.7	35.3	620,726
1984	28.5	4.9	5.6	13.3	14.1	33.6	736,972
1985	29.3	5.3	6.9	14.4	11.0	33.2	813,607
1986	34.7	4.6	6.6	18.1	6.6	29.4	827,937
1987	32.6	5.0	8.1	19.0	5.0	30.2	948,551
1988	36.5	5.0	9.5	20.2	2.8	26.0	1,111,818
1989	39.9	4.3	7.9	16.8	6.3	24.8	1,377,636
1990	44.3	4.5	6.4	16.6	5.1	23.1	1,561,479

Sources: Department of Statistics and Research, *Tourism, Migration and Travel Statistics*, *Annual Reports*, Nicosia; Cyprus Tourism Organization, *Annual Reports*, CTO, Nicosia.

Notes: 1 Sweden only between 1970 and 1976.
2 The figure of 43.3% for the Lebanon in 1976 reflects the political difficulties in that country as well as the downturn in arrivals from Cyprus's traditional markets following the Turkish military intervention in summer 1974.

tourist/day and the general increase of tourist traffic. This period coincides with the widespread introduction of package holidays and jet aircraft and this was reflected in the rising proportion of arrivals by air from 80 per cent in 1966 to 93 per cent by 1971.

In 1971, the composition of tourists by nationality showed a strong dependence on just three markets (UK, USA and Greece) which accounted for about 60 per cent of arrivals. Germany and Scandinavia contributed only 8.7 per cent and Lebanon's share was 6.0 per cent (Table 12.1). The promotion of Cyprus in West Germany and Scandinavia, though at a very early stage, was nevertheless beginning to yield positive results. The upward trend in arrivals was maintained to mid-1974 with growing proportions of arrivals from West Germany and Sweden. Most of the additional traffic was concentrated in Famagusta and Kyrenia where a major construction boom was underway (Table 12.2). These resorts accounted for 65 per cent of hotel beds and 73 per cent of arrivals in the 1973 season. In contrast, the remaining coastal districts (Limassol, Larnaca and Paphos) had witnessed little development (Table 12.3 and Figure 12.2).

Table 12.2 Hotel construction and star rating in Famagusta and Kyrenia in 1974

| | Famagusta | | Kyrenia | |
Star rating	pre-1970	1970–4	pre-1970	1970–4
4	1	11	1	1
3	3	11	2	3
2	1	8	1	7
1	2	1	4	5
Totals	7	31	8	16

Source: Cyprus Tourism Organization (1973) *Cyprus Hotels Guide 1974*, CTO, Nicosia.

In Famagusta, development was concentrated in the Greek-Cypriot suburb of Varosha (Maras) which lies south of the old walled city and port area, while a secondary group of hotels lay about 10 km to the north near the ruined city of Salamis. Construction in the Kyrenia district was more dispersed. Most of the larger hotels were built along the coast westwards from

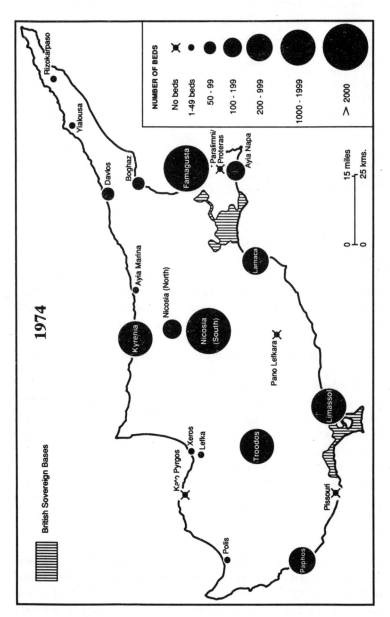

Figure 12.2 Distribution of hotels and hotel apartments in 1974

Table 12.3 Distribution of tourist accommodation in the Republic of Cyprus, June 1974

Region	Hotels 1–5 star		Hotel apartments	
	Number	*Beds*	*Number*	*Beds*
Famagusta	41	6,345	35	2,994
Kyrenia	24	1,882	31	c. 1,000
Larnaca	3	308	n/a	
Paphos	3	280	n/a	
Limassol	10	1,285	n/a	
Nicosia	19	1,038	n/a	
Hill resorts	16	1,058	n/a	

Source: Cyprus Tourism Organization (1973) *Cyprus Hotels Guide 1974*, CTO, Nicosia

Kyrenia harbour towards Vavilas (Guzelyali). Only a few hotels were built on the relatively remote north-east coast such as at Ayia Marina (Kucukerenkoy) and Dhavlos (Kaplica). More than half the hotels in Famagusta and Kyrenia were in the three- or four-star categories with a marked tendency towards larger and higher quality accommodation in Famagusta. The majority of the investment was local capital and most hotels were owner-operated though some chains, such as the Lordos properties in Varosha, were beginning to emerge. Public expenditure was concentrated on the Golden Sands complex (986 beds) in Varosha and the largest example of overseas investment was the Salamis Bay Hotel (720 beds) by the UK-based Fairclough construction group. In addition, government supplied the Pakhyamos and Alakati sites in Kyrenia with basic infrastructural utilities such as electricity, water, roads and telephone lines. A 600-bed complex was planned for the latter, and was just one of about fifty hotel and hotel apartment projects which were brought to a sudden halt during summer 1974.

The Turkish military intervention that followed the short-lived Greek-Cypriot-engineered coup resulted in the formation of a Turkish Cypriot area in northern Cyprus that amounted to 37 per cent of Cyprus and included Kyrenia, most of Famagusta District and northern Nicosia. The hostilities are fully documented elsewhere (Hitchens 1984; Birand 1985); however, in terms of the tourist industry the possession of

Table 12.4 Hotels, hotel apartments and guest houses in northern Cyprus, which came under Turkish control in summer 1974

Region	Hotels 1–5 star		Hotels without star and guesthouses		Hotel apartments		Hotels under construction		Hotel apartments under construction	
	Number	Beds	Number	Beds	Number	Beds	Number	Beds	Number	Beds
Famagusta	41	6,345	16	294	35	2,994	11	1,844	20	1,180
Nicosia	1	141	1	24	–	–	–	–	–	–
Kyrenia	24	1,882	5	82	31	c. 1,000	6	955	4	500
Morphou	–	–	1	20	–	–	–	–	–	–
Totals	66	8,368	23	420	66	c. 4,000	17	2,799	24	1,680

Sources: Cyprus Tourism Organization (1973) Cyprus Hotels Guide 1974, Nicosia; ibid., List of Hotels, Hotel Apartments and Guest houses now controlled by the Turkish Army, unpublished typescript, nd.

almost every major coastal hotel passed into the hands of the Turkish Cypriot community (Table 12.4). In contrast, Larnaca and Paphos together could muster only three hotels in the three- and four-star categories.

The military defeat of the Greek-Cypriot forces created a huge refugee problem as virtually the whole Greek-Cypriot population formerly resident in the North fled south. Moreover, Nicosia International Airport now lay in the buffer zone between the two communities and has remained closed and new airports have since been constructed at Larnaca and Paphos. About 82 per cent of tourist accommodation and 96 per cent of new hotels under construction were lost and consequently tourism faced imminent collapse. The events of 1974 vividly highlighted not only the shortcomings of tourism strategy, in failing to ensure a proper regional distribution of resort development, but also the heavy dependence of the national economy on tourism which accounted for over 40 per cent of exports in the early 1970s.

NORTH CYPRUS

The *de facto* division of the island heralded a major slump of tourism in the Turkish Cypriot region and it was not until the late 1980s that sustained growth was evident. Political, economic and military factors largely account for the failure to efficiently utilize existing facilities. On the one hand the North lacks official diplomatic recognition and, although trading with many European and Middle East states, only has diplomatic ties with Turkey, which retains a 30,000 strong military presence and several republics of the former Soviet Union. The government of the Republic of Cyprus (Greek Cypriot) has declared the airports and seaports within North Cyprus illegal points of entry so that no foreign nationals arriving at those places can subsequently travel into the Republic.

Moreover, visitors holidaying in the South can only make day visits to the North via a checkpoint in Nicosia, and even this access was closed for much of 1990 due to tension between the two communities. Civil aviation restrictions prevent direct flights between European airports and North Cyprus and aircraft are routed via Istanbul or Izmir in Turkey thus adding to journey times and increasing transport costs. A further difficulty concerns Varosha which has remained a closed military area inaccessible even to Turkish Cypriots. Only two hotels on its fringe are open while about another ninety properties, together with the business district, now lie derelict.

The demise of the tourist industry is less pronounced in the Salamis area and in Kyrenia District. The main losses were among smaller hotels and restaurants on the north coast (Dhavlos (Kaplica)) and in the back streets of Kyrenia. Many large units in various states of completion were abandoned and only a few of these were subsequently finished, largely in response to the recent upsurge of arrivals (Jasmine Court Hotel, Kyrenia).

Bed capacity slumped dramatically from over 12,700 in the 1974 season to just 2,300 in the following year. The number of tourist visitors also plunged. Famagusta and Kyrenia had attracted 186,000 visitors in 1973; however, only some 68,000 arrivals were recorded as having visited the whole of North Cyprus in 1975 and 90 per cent of these were Turkish. In contrast, tourist arrivals in 1990 amount to 300,000 while bed capacity has more than doubled. Foreign exchange earnings rose from US $30.2 million in 1977 to US $225 million in 1990.

Recovery has been a very gradual process, interrupted by a slump in arrivals about 1980 and is heavily dependent on visitors from Turkey. Without Turkish tourists, who visit throughout the year, the holiday industry in North Cyprus would have virtually disappeared. Nevertheless, dependence on Turkish visitors has been one of the industry's greatest weaknesses because many only visit for short periods, often on shopping trips, and either stay with friends or in cheaper accommodation. Diversification into other markets has been limited with Britain and Germany accounting for the majority of visitors from 'third' countries (Table 12.5).

One of the major problems hindering tourism development has been infrastructural bottle-necks, particularly in communications. The performance of Turkish Airlines and lack of liaison with ferry operators were sources of complaint in the early 1980s. However, substantial improvements have been made in recent years. A major new harbour was completed at Kyrenia in 1987 and charter air services from several UK airports now operate to Ercan Airport east of Nicosia, while

Table 12.5 Tourist arrivals staying in hotels in northern Cyprus by nationality, 1983–90 (%)

	UK	Turkey	Germany	Scandinavia	Other sources	Total arrivals
1983	7.8	72.0	4.7	0.2	15.1	49,170
1984	9.0	71.2	5.5	0.2	14.1	47,115
1985	14.2	64.4	7.0	0.4	15.3	40,030
1986	13.0	57.5	13.1	0.7	15.6	40,963
1987	11.1	60.3	13.8	2.0	12.7	78,984
1988	10.6	62.6	13.3	1.9	10.8	86,352
1989	14.7	58.0	13.0	1.9	12.5	83,250
1990	15.5	58.3	12.2	1.9	12.1	82,318

Source: Turkish Republic of Northern Cyprus, Tourism and Planning Office, *Annual Tourism Statistics*, Lefkosa.

Note: Statistics for individual countries other than Turkey are only published for tourists staying in hotels.

Saudi Arabian and Turkish aid has been used to upgrade the trunk road network. However, there were a number of even more fundamental problems that hindered tourism development, namely a lack of advertising, shortages of appropriate staff, erratic arrival patterns and the small size of most hotels which remained open. In 1990 only five hotels were in the four- and five-star categories and a similar pattern characterized the self-catering sector.

Recent seasons have, however, seen a major turn in the fortunes of tourism in the North. To some extent this may be related to the growing popularity of Turkey and southern Cyprus and it is possible that some visitors to the South subsequently choose to holiday in the North. More likely the explanation lies in the increasing amount of advertising, the wider availability of charter flights and the growing range of hotels and apartments. A further boost was the enactment in 1987 of 'The Tourism Promotion Bill' which initiated a wide range of financial incentives to entrepreneurs, hoteliers and package holiday operators. Amongst these are low rents of state land, tax concessions, foreign investors could repatriate profits, duty exemptions for imported goods and government contribution to charter risk. The policy is already bearing fruit and eighteen hotels (2,708 beds) were under construction and a further four were being expanded in 1988. Several near Kyrenia and at Salamis had already opened in time for the 1990 season.

Some development is also taking place in the Karpas peninsula and at the small hill resort of Kantara and it is official policy to encourage village tourism. Although many of these units are totally new developments, some make use of buildings that were abandoned in 1974 (Zephyros, Karavas (Alsancak)), while others have taken advantage of serviced sites where the infrastructure was put in during the early 1970s. Investors include Cyprus Turkish Tourism Enterprises, a parastatal organization formed to manage the larger Greek-Cypriot-owned hotels, and international and Turkish business interests such as Polly Peck and Noble Reardon. Already the multiplier effect is apparent in the growing number of restaurants, car hire and tour companies.

REPUBLIC OF CYPRUS

The loss of accommodation and other tourism resources in Kyrenia and Famagusta to the Turkish forces at the height of the summer season in 1974 was a major blow. In 1975 arrivals slumped to just 47,000 compared with North Cyprus which recorded 68,000 visitors. Such has been the pace of recovery that in every year since 1976 visitors to the South have exceeded those to the North and today the ratio in favour of the Republic is just over 5:1. Indeed, growth has been at a faster rate than other Mediterranean countries.

Moreover, Cyprus is not dependent on a single market for the majority of its tourists. Although the British market is the single most important source, currently accounting for about two-fifths of arrivals, Scandinavia and West Germany now contribute more than a quarter of visitors while Lebanon and Greece each have a 5 per cent share.

Rapid growth in part reflects the underdeveloped nature of tourist resources in the South, the policies of government and the parastatal Cyprus Tourism Organization (CTO). The First Emergency Plan aimed to re-establish Cyprus on the tourist map, ensuring a satisfactory occupancy rate for the remaining accommodation and the construction of ancillary facilities such as the Larnaca marina, beach development at Limassol and upgrading picnic sites. A number of major bottlenecks were identified including lack of accommodation in coastal locations, the absence of the necessary infrastructure

239

in these areas, the seasonality of the industry and the reluctance among potential tourists from some markets to visit Cyprus. It was not until the late 1970s (Second Emergency Economic Action Plan) that many of these difficulties began to be tackled.

Several of the most important policies were formulated at this time, particularly the decision to attract tourists from higher income groups, to diversify markets and to lengthen the season. The success of these measures has been one of the hallmarks of tourism development (Table 12.6). Accompanying the policies has been a series of initiatives to increase the supply of accommodation including long-term lease of State land at Ayia Napa and Paralimni, a short distance south of Famagusta, and loans channelled thorugh a Special Fund for the Financing of Priority Projects. As a result, Cyprus has largely developed as a 'sea and sun' resort with the Troodos Mountain villages providing only a minor counterbalance. Development has concentrated on four coastal locations; first

Table 12.6 Tourism indicators for the Republic of Cyprus and the Turkish Republic of Northern Cyprus, 1990

	Republic of Cyprus	*TRNC*
Net foreign exchange earnings	573.0 (CY £m)	224.8 (US $m)
Full-time employment	29,050	3,615
Number of hotels[1]	185	29
Number of hotel apartments	234	26
Bed capacity[1]	50,675	5,414
Average occupancy rate (%)		
hotels	64.3	37.6
hotel apartments	61.5	
Tourist arrivals[2]	1,561,479	300,810

Sources: Cyprus Tourism Organization, *Annual Report 1990*, Nicosia; Turkish Republic of Northern Cyprus, Tourism and Planning Office, *Annual Tourism Statistics 1990*, Lefkosa.

Notes: 1 Includes hotels in the one- to five-star categories. In 1990 there were forty-four guesthouses and hotels without a star (1,099 beds) and 7,833 beds in tourist villas, tourist apartments and furnished apartments registered with the Cyprus Tourism Organization. In Northern Cyprus there were in addition 711 beds in thirty-four guesthouses in 1990.

2 Only 88,126 tourist arrivals stayed in hotels in Northern Cyprus. The figure for all tourist accommodation was 115,374.

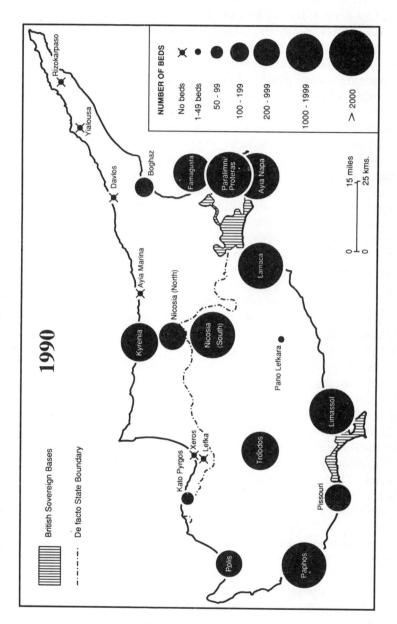

Figure 12.3 Distribution of hotels and hotel apartments in 1990

at Limassol, next Ayia Napa/Paralimni, followed by Larnaca and most recently Paphos. In each of these locations there are major clusters of hotels and hotel apartments (see Figure 12.3).

Although tourism has been a major driving force in the Cyprus economy and currently contributes 15 per cent of gross national product (GNP), a number of problems have emerged which are likely to have a bearing on its future development. There is little doubt that infrastructural improvements have not kept pace with the growth of tourism. Sewage disposal is unsatisfactory, the roads are overcrowded and there has been little provision for pedestrians in Limassol and eastern Paphos. Supporting facilities such as beaches are of an indifferent standard, many water-related sports are inadequately developed and there is only one sizeable marina. Moreover, there is a lack of product variety, holiday package costs are relatively expensive and cultural tourism does not appear to have been developed very effectively. The rapid pace of development has outstripped the planning of resources. Many of the newest hotels have encroached on beach areas, in the same way as they did in Varosha, while a plethora of new construction activities and a great deal of unlicensed accommodation puts a strain on the existing infrastructure and undermines the CTO's efforts to portray Cyprus as a quality destination.

With arrivals reaching more than 1.5 million in 1990, the authorities have begun to reappraise their tourism strategy. In particular they have come to realise that rapid growth bringing with it short-term benefits, may also herald major problems in the medium-term outlook. In June 1989 a freeze on further coastal hotel construction was imposed until October 1990. Unfortunately ancillary tourist facilities were not included and the CTO have suggested that all further development should be curbed until proper local planning schemes have been proposed. The environment lobby claim that such measures are too weak and will succumb to pressure from land developers and villagers anxious to sell their fields and give up farming. Moreover, it is argued that with a large number of building permits already granted, another construction boom will begin once the current moratorium is lifted.

In summer 1990, tourism in the Republic of Cyprus was buoyant. Arrivals from Britain continued to grow very rapidly in spite of the stagnant condition of the UK holiday market. In 1990, the most recent year for which complete statistics are available, there was an increase of 11.3 per cent in the number of arrivals, foreign exchange earnings were up by 16.9 per cent and almost 5,000 new beds were in operation.

CONCLUSION

Cyprus is a good illustration of the patterns that have characterized Mediterranean tourism development. Like many islands it has experienced rapid growth since the early 1960s and throughout the period it has continued to draw a significant proportion of its visitors from the UK. Instability arising from intercommunal violence has at various times checked growth. The late 1950s and mid-1960s were depressed times, but these were totally overshadowed by the Turkish invasion in 1974 that led to the partial closure of tourism in the North and its reconstruction in the South. Whereas the industry in the Greek Cypriot area has made a remarkable recovery, until recently the Turkish Cypriot sector has stagnated due to ineffective utilization of resources, political isolation and economic difficulties such as high rates of inflation due to the weakness of the Turkish lira. Political instability in the Middle East too has had some bearing on the pattern of arrivals. The fluctuating number of visitors from the Lebanon is the most dramatic trend and at the time of writing (March 1991) it remains to be seen whether the sharp fall in arrivals in the early months of 1991 resulting from the Gulf War will be translated into a lean period for both tourist industries. A further complication for the North relates to the financial difficulties being experienced by the Polly Peck group which has substantial interests in tourism enterprises.

The main attraction of the South to tourists of all nationalities is the 'sun and sea' image and the comparative cheapness of goods and services. Unless, however, it can upgrade its infrastructure and achieve a wider spread of facilities, it may face difficulties in the future in maintaining its share of the 'quality tourism' market. The current middle income clientele may become disenchanted with unplanned growth of ancillary tourist facilities that have created an untidy environment in Protaras, Limassol and Paphos which local commentators have likened to the 'costa syndrome'. In contrast, North Cyprus faces a different set of problems. The continuing political stalemate between the Greek Cypriot and Turkish Cypriot leadership remains the major handicap to the development of the tourist potential of the North. Dependence upon Turkish visitors, too, is a continuing difficulty and efforts to diversify arrivals have met with only limited success. Nevertheless, North Cyprus does possess certain advantages. Its proximity to the rapidly developing southern shores of Turkey does mean that two-centre holidays are feasible. Moreover, the North has the best beaches in Cyprus and these, together with its varied cultural and historical attractions, are being heavily promoted (see Figure 12.4).

In concluding, it is ironic to note that while massive construction activity has characterized the South since 1974 and efforts are being made to rejuvenate Kyrenia on the north

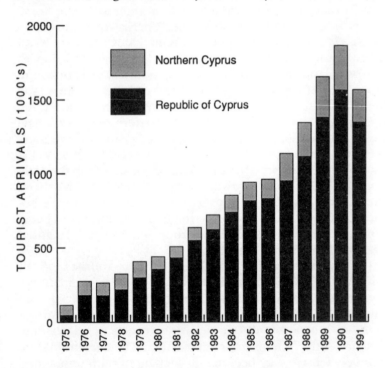

Figure 12.4 Tourist arrivals to the Republic of Cyprus and Northern Cyprus, 1979–91

coast, Varosha, once the island's premier resort, remains closed. From time to time the Turkish Cypriot leadership has threatened to reopen this area and presumably in that event some hotels might be refurbished, though given the resources available, and the degree of dereliction, it seems likely that only a few units would begin trading. On the other hand, a comprehensive solution to the political problems of Cyprus could herald the rebirth of tourism in Famagusta. Such an outcome might prove a mixed blessing as it would create an oversupply of capacity that could seriously jeopardize the operating margins of accommodation in less attractive areas of the South, such as the eastern outskirts of Limassol and Dhekalia near Larnaca, which have been developed during the last fifteen years.

NOTE

Place names in Northern Cyprus have been changed since 1974. In the text, both Greek and Turkish names are given for these places with the exception of Kyrenia, known as Girne, Famagusta (Gazimagusa), Morphou (Guzelyurt) and Nicosia (Lefkosa).

REFERENCES

Andronikou, A. (1979) 'Tourism in Cyprus', in E. de Kadt (ed.) *Tourism – Passport to Development?*, London: Oxford University Press.

Andronikou, A. (1987) *Development of Tourism in Cyprus: Harmonization of Tourism with the Environment*, Nicosia: Cosmos.

Bastin, R. (1984) 'Small island tourism: development or dependency?' *Development Policy Review* 2(1): 79–90.

Birand, M.A. (1985) *30 Hot Days*, London and Nicosia: Rustem.

Butler, R.W. (1993) 'Tourism development in small islands: past influences and future directions', in D.G. Lockhart *et al.* (eds) *The Development Process in Small Island States*, London: Routledge.

Cyprus Tourism Organization, *Annual Reports 1980–90*, Nicosia: CTO.

Economic Division Ministry of Trade and Economic Planning, Malta (1986) *Economic Survey January–September 1986*, Malta: Government Press.

English, E.P. (1986) *The Great Escape? An Examination of North-South Tourism*, Ottawa, Canada: The North-South Institute, 58–61.

Gillmor, D.A. (1989) 'Recent tourism development in Cyprus', *Geography* 74(3): 262–5.

Hitchens, C. (1984) *Cyprus*, London: Quartet.

Iktisadi Arastirmalar Vakfi (1989) *Kuzey Kibris Ekonomisinin Gelismesinde Hizmetler Sektoru, Girne 7 Aralik 1988*, Istanbul.

Lea, J. (1988) *Tourism and Development in the Third World*, London: Longman.

Lockhart, D.G. and Ashton, S.E. (1990) 'Tourism to Northern Cyprus', *Geography* 75(2): 163–7.

Lockhart, D.G. and Ashton, S.E. (1991) 'Tourism in Malta', *Scottish Geographical Magazine* 10(1): 22–32.

Matthews, J.E. (1964) *Cyprus: An Economic and Geographical Outline*, Nicosia: Zavallis Press, 2nd edn.

Ministry of Commerce and Industry, Cyprus and Affaires Etrangeres Ministry, France (1962) *Cyprus: Study of Tourist Development*, Paris.

Pearce, D.G. (1987) *Tourism Today: A Geographical Analysis*, London: Longman.

Planning Bureau, Republic of Cyprus, *The Second Five Year Plan (1966–71)*, Nicosia.

Planning Bureau, Republic of Cyprus, *The Third Five Year Plan (1972–6)*, Nicosia.

Planning Bureau, Republic of Cyprus, *Emergency Action Plan, 1975–6*, Nicosia.

Planning Bureau, Republic of Cyprus, *Second Emergency Economic Action Plan 1977–8*, Nicosia.

Planning Bureau, Republic of Cyprus, *Third Emergency Economic Action Plan, 1979–81*, Nicosia.

Planning Bureau, Republic of Cyprus, *Fourth Emergency Economic Action Plan 1982–6*, Nicosia.

Prime Ministry, State Planning Organization, Turkish Republic of Northern Cyprus (1988) *Economic and Social Developments in the Turkish Republic of Northern Cyprus*, Lefkosa.

Prime Ministry, State Planning Organization, Turkish Republic of Northern Cyprus (1988) *Second Five Year Development Plan (1988–92)*, Lefkosa.

Turkish Republic of Northern Cyprus, Tourism and Planning Office, *Tourism Statistics* (Annual), Lefkosa.

United Nations Programme of Technical Assistance (1961) *Cyprus – Suggestions for a Development Programme*, New York (Thorp Report).

Wilson, R. (1988) 'The impact of tourism on the economy of Cyprus', *Manchester Papers on Development* 4: 226–43.

World Bank (1987) *Cyprus: A Long-term Development Perspective*, Washington DC: World Bank.

13

DEVELOPMENT AND THE NATURAL ENVIRONMENT IN THE MALTESE ISLANDS

Patrick J. Schembri and Edwin Lanfranco

INTRODUCTION

The Republic of Malta is a small Mediterranean island nation with a limited land area (< 316 km^2) and very high population density (at 1,095 per km^2, the highest in Europe). As might be expected, human pressure on the natural environment is intense. Development of the Maltese islands commenced some 7,000 years ago with the arrival of the first colonists who radically modified the landscape by clearing the native forests and other natural vegetation for agriculture, construction and fuel, and by the introduction of grazing animals which prevented trees from regenerating. Other significant modifications took place over the period 1530–1798 when the islands experienced a population explosion, a building boom and heightened agricultural activity, all connected with the reign of the Knights of Saint John. Improved medical services and sanitation after the islands passed into British hands caused a second population explosion in the early twentieth century with a concurrent spread of settlements, causing further modification to the natural landscape. This was mainly caused by building activity connected both with housing and with military installations. The population has continued increasing steadily since.

This contribution describes and analyses the main effects of development on the natural environment of the Maltese islands. Three key factors shape the Maltese natural environment: the geology, the climate and man. In order to understand

247

the effect of man it is therefore necessary to have an understanding of the other two factors. Accordingly, before discussing human impact, a brief overview of the physical geography of the Maltese islands is given.

GEOGRAPHICAL BACKGROUND

The following account is based on the works of: Murray (1890), Bowen-Jones *et al.* (1961) and Ransley (1982) on general geography; Hyde (1955), House *et al.* (1961), Pedley *et al.* (1976, 1978) and Zammit Maempel (1977) on geology; Lang (1960) on soils; House *et al.* (1961) and Vossmerbäumer (1972) on geomorphology; Reuther (1984) on tectonics; Mitchell (1961) and Chetcuti (1988) on climate; Sommier and Caruana Gatto (1915), Borg (1927), Haslam (1969), Haslam *et al.* (1977) and Lanfranco (1989) on vegetation; and Haslam (1969) and Schembri (1988) on general ecology.

Geology and geomorphology

The Maltese islands are a group of small, low islands aligned in a NW-SE direction and located in the central Mediterranean. The Maltese archipelago consists of three inhabited islands: Malta, Gozo and Comino and a number of small uninhabited islets.

Geologically, the islands are composed almost entirely of marine sedimentary rocks, mainly limestones of Oligo-Miocene age, capped by minor Quaternary deposits of terrestrial origin. The five main rock types are (in order of decreasing age): Lower Coralline Limestone, Globigerina Limestone, Blue Clay, Greensand, and Upper Coralline Limestone. Maltese soils are characterized by their close similarity to the parent rock material, their relatively young age, the ineffectiveness of the climate in producing soil horizon development, and the great importance of human activities in modifying them.

Erosion of the different rock types gives a characteristic topography. Lower Coralline Limestone forms sheer cliffs which bound the islands to the west; inland this rock type forms barren, grey, limestone-pavement plateaux on which karstland develops. The Globigerina Limestone, which is the most extensive exposed formation, forms a broad rolling landscape. Blue Clay slumps out from exposed faces to form taluses of about 45° over the underlying rock. Upper Coralline Limestone forms massive cliffs and limestone pavements with karstic topography similar to the Lower Coralline Limestone.

Both main islands are tilted seawards to the northeast. There are no mountains, the highest point on Malta is only 253m above sea level; the highest point on Gozo is 191m. There are also no lakes, rivers or streams, only minor springs.

Characteristic topographic features of particular ecological importance are the *rdum* and *widien* (singular *wied*). *Rdum* are near-vertical faces of rock formed either by erosion or by tectonic movements. Their bases are invariably surrounded by screes of boulders eroded from the *rdum* edges. Because of the shelter they provide and their relative inaccessibility, the *rdum* sides and boulder screes provided important refuges for many species of Maltese flora and fauna, including many endemics. *Widien* are drainage channels formed either by stream erosion during a previous (Pleistocene) much wetter climatic regime, or by tectonism, or by a combination of the two processes. Most *widien* now carry water along their watercourses only during the wet season, a few, however, drain perennial springs and have some water flowing through them throughout the year. By virtue of the shelter provided by their sides and their water supply, *widien* are one of the richest habitats on the islands.

Climate

The Maltese climate is typically Mediterranean. The average annual precipitation is just over 500mm. Rainfall is highly variable from year to year; some years are excessively wet while others are extremely dry (extreme minimum for period 1854–1986, 191.3mm; extreme maximum for period, 1,031.2mm). The seasonal distribution of rainfall defines a wet period (October to March with about 70 per cent of the total annual precipitation) and a dry period (April to September). Air temperatures are moderate (mean annual temperature for period 1951–86, 18.5°C; mean monthly range, 12.3–26.2°C) and never fall too low for adequate plant growth. Evapotranspiration is high and accounts for between 70–80 per cent of the total annual precipitation. Wind blows on approximately 87 per cent of the days of the year. Only some 16 per cent of the rainfall infiltrates into the substratum and can be used by vegetation for growth. Summary climatic data are given in Table 13.1.

Table 13.1 Maltese islands: mean monthly values for selected climatic parameters (calculated for period 1951–80)

Month	Rainfall (mm)	Temperature (°C) (max.)	(min.)	Sea temp. (°C)	Sunshine (hours)	EVT (mm)
Jan.	88.2	15.0	9.5	14.5	5.3	28
Feb.	61.4	15.4	9.4	14.5	6.3	28
Mar.	44.0	16.7	10.2	14.5	7.3	35
Apr.	27.5	18.7	11.8	16.1	8.3	44
May	9.7	23.0	14.9	18.4	10.0	70
Jun.	3.4	27.4	18.6	21.1	11.2	100
Jul.	0.9	30.2	21.0	24.5	12.1	132
Aug.	9.3	30.6	21.8	25.6	11.3	135
Sep.	44.4	27.7	20.2	25.0	8.9	105
Oct.	117.9	23.7	17.1	22.2	7.3	81
Nov.	75.5	19.9	13.8	19.5	6.3	55
Dec.	96.0	16.7	11.1	16.7	5.2	38

EVT = mean monthly evapotranspiration estimated by the method of Thornwaite.

Habitats and biota

There are indications that originally much of the Maltese islands were covered by climax Mediterranean sclerophyll forest, dominated by Holm Oak (*Quercus ilex*) and Aleppo Pine (*Pinus halepensis*), with maquis scrubland and garigue communities in places where edaphic factors and exposure prevented climax forest from developing. Much of the natural vegetation was cleared by the early neolithic settlers to provide land for agriculture and habitation. Currently there are only remnants (occupying only a few tens of square metres) of the original climax woodland. Other woodland areas on the islands, the most important of which is Buskett in southwest Malta, have been planted by man in relatively recent times.

Maquis communities occur in small patches in sheltered situations as, for example, on the sides of the deeper *widien*, at the bases of *rdum* and amongst the boulder screes surrounding them. Maquis also develops in very small patches of a few square metres in association with Carob (*Ceratonia siliqua*) and Olive (*Olea europaea*) trees planted since at least classical times round the periphery of fields. Hilltops, cliff verges, the edges of *widien*, coastal karstland and other exposed ground support garigue communities of small perennial bushes, geophytes and annuals. Where this has been degraded, steppe communities dominated by grasses (Gramineae), umbellifers (Umbelliferae), thistles and Asphodel (*Asphodelus aestivuus*) develop. Garigue and steppe are the most widespread natural vegetational communities on the Maltese islands and show many different subtypes, amongst which is a maritime garigue/steppe dominated by halophytes and xerophytes. Erosion of the Blue Clay produces clay slopes which support a distinctive vegetation dominated by grasses of which the most important is Esparto Grass (*Lygeum spartum*). Other minor habitats include freshwater marshlands, saline marshlands, sand dunes, rainwater pools, cliffsides and caves.

Additionally, human activities have created a variety of habitats such as fields, gardens, road verges and land cleared of the natural vegetation cover for a variety of purposes. Of these, those not under active management become invaded by a flora of weed species.

In spite of their limited land area and habitat diversity, and the intensive human pressure on the natural environment, the Maltese islands support a rich and diverse biota, certain elements of which are of particular scientific and cultural importance. For example, there are some 1,000 species of flowering plants and an equal number of lower plants, some fifty species of freshwater and terrestrial molluscs, more than 4,000 species of insects, one amphibian, nine terrestrial reptiles, some thirteen resident, fifty-seven regularly visiting and 112 migrant birds, and some twenty-one species of mammals. A relatively large number of species of plants and animals are found in the Maltese islands only and nowhere else in the world. These include some twenty-one flowering plants, seventeen molluscs, seventeen moths and butterflies, more than twenty-five beetles, some twenty other species of invertebrates and one reptile.

HUMAN IMPACT

The main episodes of Maltese history have been summarized by Blouet (1984). At present the islands' total population is 345,418 (1985 census) distributed as follows: Malta 319,736, Gozo 25,670 and Comino 12.

Loss of countryside

Development of the Maltese islands has been particularly rapid during the past thirty years. Land area occupied by buildings increased from 5 per cent in 1957 to 16 per cent in 1985 (Ministry for Development of Infrastructure 1988; see Figure 13.1).

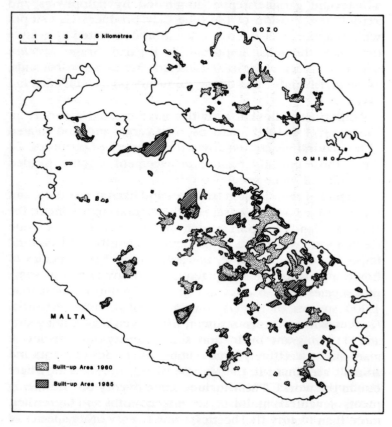

Figure 13.1 Urban expansion between 1960 and 1985

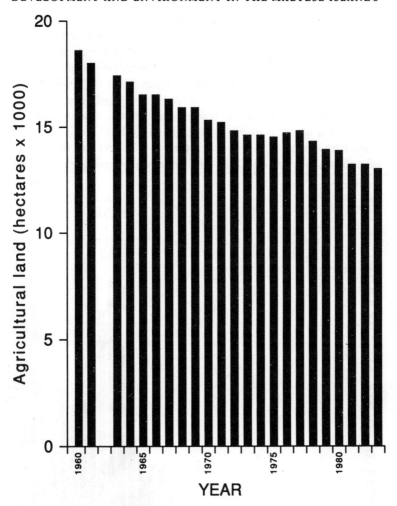

Figure 13.2 Registered agricultural land in the Maltese islands 1960–83

Concurrently, registered agricultural land decreased from c. 177 km^2 (56 per cent) in 1957 to c. 120 km^2 (38 per cent) in 1983 (Central Office of Statistics) (Figure 13.2), and quarrying activity increased to supply the growing demand for building material. The main effects of this has been the replacement of natural landscapes by anthropic ones and, consequently, the loss of habitats and concurrent reduction in wildlife populations.

Another aspect of this intense development was increased road construction. For example, the 893 km of roads existing in 1957 became 1,482 km in 1987 (Central Office of Statistics), a 66 per cent increase. More significant in terms of environmental impact, however, is the construction of access roads to previously remote areas which have encouraged new settlements away from the traditional centres. The deposition of rubble and the dumping which invariably accompany road construction has modified the countryside to the detriment of habitats and biota. This problem has been aggravated by

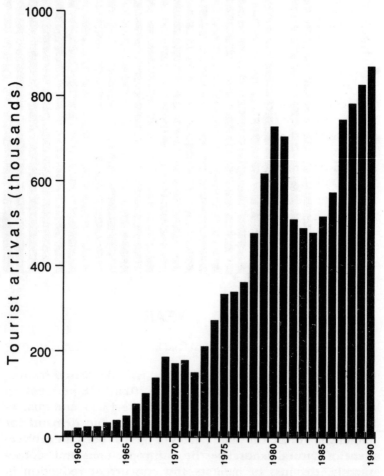

Figure 13.3 Tourist arrivals 1957–90

the increased number of motor vehicles on the roads, rising from 15,929 in 1957 to 117,150 in 1987 (Central Office of Statistics), a 635.5 per cent increase. Problems caused to the natural environment are a consequence of the toxic components of exhaust and of littering and dumping.

In the 1950s the islands were opened to tourism which, after a modest start, reached a peak in 1980 with almost 30,000 arrivals. The number of tourists declined to 480,000 in 1984 and then peaked again at 871,776 in 1990 (Central Office of Statistics) (Figure 13.3). Development connected with the tourist industry has eaten up more of the countryside, particularly in coastal areas (Anderson and Schembri 1989).

Waste disposal

After consumption of the limited land resources, perhaps the most serious environmental problem is the disposal of waste generated by the permanent and transient population and by industry.

There are three official dumping sites on Malta and one on Gozo. Although the material in these dumps is frequently burnt, either deliberately or accidentally, and reduces in bulk, all four sites are expanding. Apart from being visually offensive, these dumps consume the land and produce clouds of smoke and noxious emissions. Additionally, a large number of unofficial dumps exist. These sites are cleared periodically by the Public Cleansing Division. However, rubbish soon accumulates again, sometimes overnight.

A large amount of rubbish is unofficially dumped into the sea. The extent of this problem is not known since, unlike terrestrial dumps, marine rubbish dumps are generally not visible. However, sports divers report quite serious accumulations in some areas.

Some sewage is processed by a sewage treatment plant. However, most still ends up in the sea. Apart from natural organics, this sewage includes various household and industrial pollutants. This has resulted in considerable degradation and modification of coastal marine biotic communities. For example, in some areas algal communities dominated by brown seaweeds (Phaeophyta), mainly species of *Cystoseira*, have been replaced by others dominated by nitrophilous green seaweeds (Chlorophyta), mainly ulvoids.

Quarrying

Globigerina Limestone is quarried for use as a building stone while Coralline Limestone is quarried for use as spalls. Many of the old quarries have been worked out and abandoned without any reclamation, while new quarries are being established. The area of land quarried is therefore increasing. Some of the new quarries are situated in ecologically sensitive areas. For example, two quarries on the southern coast of Malta have broken completely through the coastal cliffs, a quarry on the south-east coast of Gozo has developed at the expense of the coastline, while a quarry at Wied il-Ghasel in central Malta, is threatening to cause the collapse of part of the valley side. The extensive quarrying in the Wied il-Ghasel/San Pawl tat-Targa area has already caused the extinction of some important species, for example, the Siculo-Maltese endemic orchid *Ophrys oxyrrhynchos* (Lanfranco 1989).

Deforestation and afforestation

Deforestation caused by the first settlers on the islands and their domestic animals has already been mentioned. There is evidence that more trees were lost during the Middle Ages. During that period the islands were frequently raided by corsairs operating from the north coast of Africa, and were at times severely depopulated (Blouet 1984). This led to much previously-cultivated land being abandoned with resultant loss of soil and the tree cover. During the reign of the Knights of St John agriculture thrived and many trees were planted for agricultural, commercial, recreational and decorative purposes. The semi-natural woodland at Buskett originates from this period when the natural woodland there was heavily augmented with other trees to turn the area into gardens and a hunting park (Borg 1979). Extensive deforestation occurred during the 1914–18 war when the islands were blockaded by enemy ships and large numbers of trees were felled for firewood. More trees were felled when the Luqa airfield was constructed and extended to connect with that at Hal Far.

At present only very small remnants of the original native forest remain at four localities (Schembri *et al* 1987; Lanfranco 1989). Each of these occupies an area of a few tens of square metres only. All other present-day wooded areas have been planted by man in relatively recent times. A drive to afforest the islands started in the early 1950s and has continued to the present time. Afforestation projects have often been attempted in unsuitable areas. Thus, following the recommendation of several reports (Hamilton 1952; Keith 1956a,b; 1963), in the late 1950s and early 1960s, the Coralline Limestone plateaux of the Marfa, Bajda and Mellieha Ridges were planted with trees after blasting holes in the rock to accommodate soil. Results from these projects have been mixed, with many trees dying or growing in a stunted condition, but others surviving to give the present wooded areas of Mizieb and Marfa Ridge. However, even though these areas are wooded, they are not woodlands as is Buskett, for example. The bare rock and shallow soil pockets between the blasted tree-holes prevent woodland-type undergrowth from developing. Additionally, the presence of trees changes the character of the physical environment resulting in the elimination of the more specialized indigenous plants and of their associated fauna.

Alien species are more commonly utilized in afforestation projects than are indigenous and archaeophytic species. Until recently, Acacias (*Acacia* spp.) were indiscriminately planted in all manner of localities, including in ecologically sensitive areas where they tend to displace native species. Eucalyptus (*Eucalyptus* spp.) have also been extensively used. These trees prevent most other plants from growing in their vicinity due to allelopathic effects.

Soil erosion, grazing and land reclamation

In the north of Malta large bare rocky areas are still enclosed by long lines of rubble walls, showing that these areas were once covered with soil and cultivated. These areas lost all their soil a long time ago, mainly because, following cessation of cultivation, the land was given over to grazing, which removed the protective vegetation cover. Without such grazing, uncultivated land would, if adequately sited, generate a maquis vegetation.

Much agricultural land is on sloping ground which is terraced with retaining walls made of limestone rubble. Many of these rubble walls have fallen into disrepair with a concurrent increase in soil erosion. Most of the agricultural land is not irrigated, which leaves the soil bare of vegetation during the dry period of the year, leading to accelerated erosion. Additionally, during the transition from the dry to the wet season, short but very heavy rainstorms are common, which lead to increased runoff and erosion. Loss of soil through runoff is accentuated due to the large number of roads which provide an unimpeded channel to the sea for storm water.

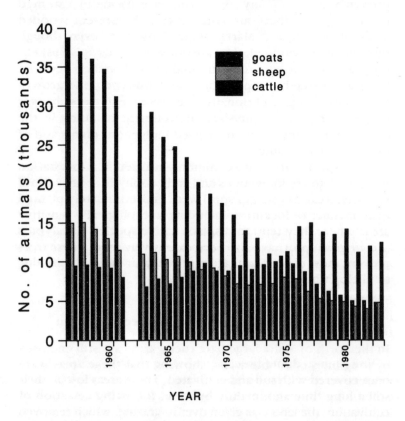

Figure 13.4 The number of goats, sheep and cattle, 1957–83

Grazing, especially by domestic goats and, to a lesser extent, sheep, has been an important factor in shaping the Maltese landscape. Land considered unfit for cultivation, or in excess of needs at the time, was given over to grazing. Until recently, large herds roamed the islands, with the result of making the land unsuited for the development of maquis and woodland vegetation. The effects of grazing were compounded by burning, often to clear 'weeds'. These pressures have resulted in garigues and steppes being the dominant vegetation types in the Maltese islands. Recently the upkeep of herds of goats and sheep has become uneconomical and breeders have been encouraged to invest in cattle. This has led to a sharp decrease in the populations of caprines and ovines (Figure 13.4) with the result that some areas are now starting a slow process of maquis regeneration, for example, in the southern and south-eastern parts of Malta.

Atempts have been made at reclaiming land by covering so called 'barren' tracts (actually karstland with typical garigue vegetation) at certain localities with soil for agricultural development. Apart from their ecological interest, many of these are very exposed raising doubt as to the long-term results of this venture.

Loss of habitats

Sandy beaches which constitute only 2.4 per cent of the islands' coastline, are under high pressure due to their recreational value (Anderson and Schembri 1989). Most of the local population makes heavy use of the coasts during the hot summer months. Tourism adds to this pressure, both directly due to use of beaches by tourists and indirectly due to the building of tourist facilities on the coasts. Building development across valleys lying behind sandy beaches, as well as the building of quays and other structures in bays, in some cases have resulted in erosion of the sand.

At present, dune ecosystems or remnants thereof exist in thirteen localities in the Maltese islands (Anderson and Schembri 1989). However, only three localities in Malta, one locality in Gozo and one locality in Comino support dunes with a more or less full dune flora, whereas the other localities support only much degraded remnants of dunes and some of these are expected to disappear completely within a few years (Schembri *et al.* 1987; Anderson and Schembri 1989). In the recent past, four previously flourishing dune ecosystems in Malta have lost all traces of the typical dune vegetation.

The important sand-binding Marram Grass (*Ammophila littoralis*) has been completely eradicated from all known sites in the Maltese islands, while the Sand Bindweed (*Calystegia soldanella*) disappeared within the last 50–60 years (Lanfranco 1989). Abusive constructions, especially in sites along the Marfa Peninsula in the north-eastern extremity of Malta, have played a role in sand-dune degradation, but a 'helping hand' has also been given from official quarters in misguided attempts to 'enhance' the amenity value of the beaches concerned. Thus, the aforementioned Marram Grass was bulldozed from its last remaining site at Ramla tat-Torri to make room for the planting of Acacias, while in the same locality the spiny *Echinophora spinosa* and Sea Holly (*Eryngium maritimum*) are regularly weeded out to make the sand suitable for sunbathing (Lanfranco 1989).

Saline marshlands are very scarce in the Maltese islands (Schembri *et al.* 1987; Anderson and Schembri 1989). Some have been completely obliterated (most recently, five sites in Malta and two sites in Gozo) and others much degraded (eight sites in Malta, three sites in Gozo, one site in Comino) by human activities (Anderson and Schembri 1989). A few are still extant (five sites in Malta), although under constant threat, while one, that at Ghadira, is a protected area (legally, a bird sanctuary), which, however, functions as a nature reserve.

Water seepage from the perched aquifers wherever the Upper Coralline Limestone/Blue Clay interface is exposed give rise to so-called high level springs which drain into *widien* watercourses. Many of these springs flowed all year round, albeit with much reduced flow in the dry period. However, most are now tapped at source by farmers to provide water for irrigation while the Water Works Department also taps these aquifers at exposed Upper Coralline Limestone/Blue Clay interfaces, further reducing the source of supply. This reduction in the number and rate of flow of high level springs has resulted in a loss of the habitat they provided (Lanfranco and Schembri 1976).

Another factor endangering local wetlands is the injudicious use of herbicides. This has become important only recently. For example, in 1984 excess herbicide used in fields bordering Bahrija Valley, one of the few sites on the islands with a permanent spring and thus of great ecological importance, killed or stunted much of the vegetation, which is only now recovering.

A recent programme of valley clearing and deepening and of small dam construction across *widien* watercourses, aimed at reducing flow along these and at retaining water in the *widien* for longer periods to allow increased infiltration and to supply water for irrigation, has provided additional freshwater habitats in the pools that form behind the dams but has also caused extensive disturbance and degradation of the *wied* ecosystems.

The ever-increasing disturbance of the natural landscape to accommodate development has resulted in a constantly changing environment which favours the spread of weed species of flora at the expense of indigenous plants. For example, the Tree Mallow (*Lavcatera arborea*) which was once uncommon, has now spread to all coastal localities. Likewise, *Aster squamatus*, which was only introduced in the 1930s, *Nicotiana glauca*, castor oil (*Ricinus communis*), and yellow wall-rocket (*Diplotaxis tenuifolia*), all of which are adventive, have now spread to all parts of the islands. The cape sorrel (*Oxalis pescaprae*) which was introduced in the early nineteenth century, has overrun the whole islands and has spread beyond the Maltese Islands to colonize much of the Mediterranean area (Henslow 1891). Some very recent introductions such as *Conyza albida* and *Solanum* cf. *americanum*, both first noted in the 1970s, are also gaining ground. However, the spread of these weed species is not an unmitigated ill since they provide a cosmetic blanket on sites which have been degraded.

Loss of wildlife

The Maltese islands support a rich and diverse biota, certain elements of which are of particular scientific and cultural importance (Schembri and Sultana 1989). A large number of species, endemic forms included, are threatened with extinction from the Maltese islands while some have already become extinct. Apart from the examples already cited, one particularly illustrative case is the draining of the marshlands at Marsa, a process which started in 1866. Among the casualties are *Sparganium erectum*, *Anthoxanthum odoratum*, *Alopecurus pratensis* and *Juncus effusus*, all plants which used to grow at Marsa but which are now extinct in the Maltese islands. Continuous interference with *widien* watercourses has resulted in the disappearance of *Ophrys apifera*, *Cyperus distachyus*, *Juncus capitatus*, *Iris pseudacorus* and others, while a number of other species of flora are gravely endangered (Lanfranco 1989). A summary of the extinct and threatened biota of the Maltese islands is given in Table 13.2.

Table 13.2 The number of extinct and threatened species of Maltese biota

Group	X	E	V	R	I
Tracheophyta (higher plants)	80	55	21	100	5
Bryophyta (mosses and relatives)	0	0	0	33	3
Crustacea (crustaceans)	0	2	2	8	2
Mollusca (snails and relatives)	2	11	7	5	2
Odonata (dragonflies and re.)	0	0	1	1	0
Dictyoptera (mantises and rel.)	0	0	0	2	0
Orthoptera (grasshoppers and rel.)	1	0	2	5	0
Dermaptera (earwigs)	0	0	1	0	0
Hemiptera (bugs)	0	0	0	1	4
Trichoptera (caddisflies)	0	0	0	2	0
Hymenoptera (bees, wasps and ants)	0	1	6	5	0
Lepidoptera (butterflies and moths)	7	1	11	9	4
Coleoptera (beetles)	11	0	37	64	48
Amphibia (amphibians)	0	0	1	0	0
Reptilia (reptiles)	0	0	11	0	0
Aves (birds)	0	10	9	2	0
Mammalia (mammals)	0	0	7	6	3

Source: Schembri and Sultana 1989.

Note: Only those groups for which reliable data exist are included and only terrestrial and freshwater forms are considered. The status classification used is the same as that employed by the International Union for the Conservation of Nature and Natural Resources (IUCN) in its *Red Data Books*: X = extinct, E = endangered, V = vulnerable, R = rare, and I = indeterminate.

CONCLUSION

As evident from the foregoing discussion, the Maltese natural environment has been subject to great pressures from a variety of development-related activities. This is understandable given its size, population density and standard of living. What was avoidable were the many mistakes in planning resulting primarily from ignorance of ecological principles and an underestimation of the value of a healthy natural environment. As a consequence, natural landscapes have been replaced by anthropic ones and there has been a general degradation of the natural environment coupled with an increase in the extent of disturbed habitats characterized by adventive species, often in competition with the indigenous biota.

It is only recently that an active awareness has been created for the need of protecting and conserving the natural environment, and to control development. The process of kindling such awareness was spearheaded by several non-governmental organizations, starting in the early 1960s. Initially having little impact on policy-makers, today there is official recognition that environmental problems exist and that they have to be tackled. Recent initiatives by government in this respect are the preparation of a Structure Plan for the Maltese islands (Ministry for Development of Infrastructure 1988), a comprehensive bill on environmental protection (Department of Information 1990), the publication of scientific appraisals of the state of certain environmental resources (Schembri *et al.* 1987; Schembri and Sultana 1989) and the accession of Malta to a number of international environmental conventions: in January 1988 Malta acceded to the 1982 Geneva Protocol on Mediterranean Specially Protected Areas; in September of the same year it acceded to the 1971 Ramsar Convention on Wetlands of International Importance, Especially as Waterfowl Habitats; and in April 1989 it acceded to the 1973 Washington Convention on International Trade in Endangered Species of Wild Fauna and Flora (CITES).

A more intelligent and environmentally-sensitive management of the natural resources of the Maltese islands will hopefully result in an acceptable blending of developmental needs and the preservation and conservation of the natural heritage of the country.

ACKNOWLEDGEMENTS

The authors are indebted to the following persons and agencies for information suplied and other help: Alfred E. Baldacchino, Deborah Chetcuti, Victor Ellul, Guido Lanfranco, John A. Schembri, Stephen Schembri, Joe Sultana, Martin A. Thake, Liz Vella, Frank Ventura, the Planning Services Division and Water Resources Management Division of the Ministry for Development of Infrastructure, and the Secretariat for the Environment of the Ministry of Education and the Interior.

REFERENCES

Anderson, E.W. and Schembri, P.J. (1989) *Coastal Zone Survey of the Maltese Islands Report*, Beltissebh, Malta: Planning Services Division, Works Department.

Blouet, B. (1984) *The story of Malta*, Valletta, Malta: Progress Press.

Borg, J. (1927) *Descriptive Flora of the Maltese Islands*, Valletta, Malta: Government Printing Office.,

Borg, J. (1979) *The Public Gardens and Groves of Malta and Gozo*, Malta: Men of the Trees (Malta).

Bowen-Jones, H., Dewdney, J.C. and Fisher, W.B. (eds) (1961) *Malta, a Background for Development*, Durham: Durham University Press.

Central Office of Statistics (annual) *Annual Abstract of Statistics*, Valletta, Malta: Central Office of Statistics.

Chetcuti, D. (1988) *The Climate of the Maltese Islands*, unpublished BEd (Hons) dissertation, Faculty of Education, University of Malta, Msida, Malta.

Department of Information (1990) *Proposed Draft Bill on Environment Protection*, Valletta, Malta: Department of Information.

Hamilton, A.P.F. (1952) *Report on a Survey of the Need for Afforestation and Water Conservation in the Wasteland of the Maltese Islands*, Valletta, Malta: Ministry of Posts and Agriculture.

Haslam, S.M. (1969) *Malta's Plant Life*, Malta: Haslam.

Haslam, S.M., Sell, P.D. and Worsley, P.A.W. (1977) *A Flora of the Maltese Islands*, Msida, Malta: Malta University Press.

Henslow, G. (1891) 'On the northern distribution of *Oxalis cernua* Thunb.', *Proceedings of the Linnaean Society 1890-1*: 31-6.

House, M.R., Dunham, K.C. and Wigglesworth, J.C. (1961) 'Geology and structure of the Maltese Islands', H. Bowen-Jones, J.C. Dewdney and W.B. Fisher (eds) *Malta, a Background for Development*, Durham: Durham University Press, 25-47.

Hyde, H.P.T. (1955) *Geology of the Maltese Islands*, Malta.

Keith, H.G. (1956a) *Draft Report to the Government of Malta on the Afforestation of Waste Lands and the Development of Tree Planting on Agricultural Lands in the Maltese Islands*, Rome: Food and Agriculture Organization of the United Nations.

Keith, H.G. (1956b) *Report to the Government of Malta on the Afforestation of Waste Lands and the Development of Tree Planting on Agricultural Lands in the Maltese Islands*, Valletta, Malta: Ministry of Posts and Agriculture.

Keith, H.G. (1963) *Report to the Government of Malta on the Progress of Afforestation on Waste Lands in the Maltese Islands* (Report no. 1724) Rome: Food and Agriculture Organization of the United Nations.

Lang, D.M. (1960) *Soils of Malta and Gozo*, London: Her Majesty's Stationery Office.

Lanfranco, E. (1989) 'The flora', in P.J. Schembri and J. Sultana (eds) *Red Data Book for the Maltese Islands*, Valletta, Malta: Department of Information, 5–70.

Lanfranco, E. and Schembri, P.J. (1986) *Maltese Wetlands and Wetland Biota*, Valletta, Malta: Society for the Study and Conservation of Nature.

Ministry for Development of Infrastructure (1988) *Structure Plan Brief*, Beltissebh, Malta: Town Planning Division, Ministry for Development of Infrastructure.

Mitchell, P.K. (1961) 'The Maltese climate and weather', in H. Bowen-Jones, J.C. Dewdney and W.B. Fisher (eds) *Malta, a Background for Development*, Durham: Durham University Press, 48–82.

Murray, J. (1890) 'The Maltese islands, with special reference to their geological structure', *Scottish Geographical Magazine* 6: 449–88 (+ plates I and II).

Pedley, H.M., House, M.R. and Waugh, B. (1976) 'The geology of Malta and Gozo', *Proceedings of the Biologists' Association* 87: 325–41.

Pedley, H.M., House, M.R. and Waugh, B. (1978) 'The geology of the Pelagian Block: the Maltese islands', in A.E.M. Nairn , W.H. Kanes and F.G. Stehli (eds) *The Ocean Basins and Margins*, vol 4B, *The Western Mediterranean*, London: Plenum Press, 417–33.

Ransley, N. (1982) *A Geography of the Maltese Islands*, Birkirkara, Malta: St Aloysius College Publications.

Reuther, C.D. (1984) 'Tectonics of the Maltese islands', *Centro* (Malta) 1: 1–20.

Schembri, P.J. (1988) *IUCN Islands Directory: Entry for the Maltese Islands* (report prepared for the IUCN Islands Directory project to be published in abridged form in the UNEP-FAO series of Directories and Bibliographies 1989).

Schembri, P.J., Lanfranco, E., Farrugia, P., Schembri, S. and Sultana, J. (1987) *Localities with Conservation Value in the Maltese Islands*, Beltissebh, Malta: Environment Division, Ministry of Education.

Schembri, P.J. and Sultana, J. (eds) (1989) *Red Data list for the Maltese Islands*, Valletta, Malta: Department of Information.

Sommier, S. and Caruana Gatto, A. (1915) *Flora Melitensis Nova*, Firenze: Stab. Pellas.

Vossmerbäumer, H. (1972) 'Malta, ein Beitrag zur Geologie und Geomorphologie des Zentralmediterranen Raumes', *Würzburger Geogr. Arb.* 38: 1–213.

Zammit Maempel, G. (1977) *An Outline of Maltese Geology*, Malta: Zammit Maempel.

INDEX